Furan Derivatives - Recent Advances and Applications

Edited by Anish Khan,
Mohammed Muzibur Rahman, M. Ramesh,
Salman Ahmad Khan
and Abdullah Mohammed Ahmed Asiri

Published in London, United Kingdom

IntechOpen

Supporting open minds since 2005

Furan Derivatives - Recent Advances and Applications
http://dx.doi.org/10.5772/intechopen.95169
Edited by Anish Khan, Mohammed Muzibur Rahman, M. Ramesh, Salman Ahmad Khan and Abdullah Mohammed Ahmed Asiri

Contributors
Kazeem Adelani Alabi, Rasheed Adewale Adigun, Mariam Dasola Adeoye, Ibrahim Olasegun Abdulsalami, El Houssine Mabrouk, Yardily A, Fathima Shahana, Abbs Fen Reji, George Kavulavu Ngusale, Laxmisha M. Sridhar, Andrew T. Slark, James A. Wilson, Antonio Pizzi, Anish Khan, Do Hyun Ryu, Dong Guk Nam, Jung Woon Yang

First published in London, United Kingdom, 2022 by IntechOpen
IntechOpen is the global imprint of INTECHOPEN LIMITED, registered in England and Wales, registration number: 11086078, 5 Princes Gate Court, London, SW7 2QJ, United Kingdom
Printed in Croatia

British Library Cataloguing-in-Publication Data
A catalogue record for this book is available from the British Library

Additional hard and PDF copies can be obtained from orders@intechopen.com

Furan Derivatives - Recent Advances and Applications
Edited by Anish Khan, Mohammed Muzibur Rahman, M. Ramesh, Salman Ahmad Khan and Abdullah Mohammed Ahmed Asiri
p. cm.
Print ISBN 978-1-83969-707-4
Online ISBN 978-1-83969-708-1
eBook (PDF) ISBN 978-1-83969-709-8

For EU product safety concerns:
IN TECH d.o.o., Prolaz Marije Krucifikse Kozulić 3, 51000 Rijeka, Croatia, info@intechopen.com or visit our website at intechopen.com.

Meet the editors

Dr. Anish Khan is an assistant professor at the Center of Excellence for Advanced Materials Research (CEAMR), Faculty of Science, King Abdulaziz University, Jeddah, Saudi Arabia. He obtained a Ph.D. from Aligarh Muslim University, India, in 2010, and a postdoctoral degree in Electroanalytical Chemistry from the School of Chemical Sciences, University Sains Malaysia (USM), in 2010. Dr. Khan works in the fields of sensors, polymer composites, and organic-inorganic electrically conducting nanocomposites. He has more than 250 research articles, 80 book chapters, 50 edited books, and more than 20 international conferences/workshops to his credit. He is an editorial board member for several international journals.

Prof. Mohammed Muzibur Rahman received his BSc and MSc from Shahjalal University of Science & Technology, Sylhet, Bangladesh, in 1999 and 2001, respectively. He received his Ph.D. from Chonbuk National University, South Korea, in 2007. He worked as a postdoctoral fellow and assistant professor in pioneering research centers and universities located in South Korea, Japan, and Saudi Arabia. Presently, he is an associate professor at the Center of Excellence for Advanced Materials Research (CEAMR) and Chemistry Department, King Abdulaziz University, Jeddah, Saudi Arabia. He has published more than 245 international and domestic conferences and several book chapters. He has also edited ten books. His research interests include photocatalysis, semiconductors, nanoparticles, carbon nanotubes, nanotechnology, electrocatalysis, sensors, ionic liquids, surface chemistry, electrochemistry, and nanomaterials.

Dr. M. Ramesh received an ME in Computer-Aided Design (CAD) from the University of Madras, Tamil Nadu, India, and a BE in Mechanical Engineering from Bharathiar University, Tamil Nadu, India. He obtained a Ph.D. in Mechanical Engineering in the field of ecofriendly composite materials from Jawaharlal Nehru Technological University (JNTU), Andhra Pradesh, India. He has published thirty-four research papers in SCI and Scopus-indexed journals, ten book chapters, and presented several papers at both national and international conferences. He has presented his research papers at universities throughout India, including IIT Guwahati, IIT Madras, University of Petroleum and Energy Studies Dehradun, Nirma University Ahamadabad, NIT Suratkal, VIT University, and others.

Prof. Salman Ahmad Khan received his master's degree in Organic Chemistry from Dr. Rammanohar Lohia Avadh University, Uttar Pradesh, India. He obtained a Ph.D. from Jamia Millia Islamia, New Delhi, India, in 2007. In the same year, he joined the Department of Chemistry, Punjabi University as a research associate. In 2008, he was an assistant professor in the Department of Chemistry, Integral University, Lucknow-UP. Dr. Khan joined the Department of Chemistry, King Abdulaziz University, Jeddah, Saudi

Arabia, in 2009 as an assistant professor and was promoted to associate professor in 2014. Recently, he joined the School of Sciences, Maulana Azad National Urdu University (MANUU) as a professor. His current research interests include heterocyclic chemistry, chromones, chalcones, Anthraquinones, cholesterol, stigmasterol, photophysical, physicochemical, and multi-step reactions, one-pot multicomponent synthesis, sensors, and nanotechnology. Dr. Salman has supervised master's and Ph.D. theses. He is also an active researcher and has published more than 150 research articles in international journals.

Prof. Abdullah Mohammed Ahmed Asiri is a professor and chairman of the Chemistry Department, King Abdulaziz University, Jeddah, Saudi Arabia. He is also the director of the university's Center of Excellence for Advanced Materials Research (CEAMR). He obtained a Ph.D. in tribochromic compounds and their applications from the University of Wales College, Cardiff, UK, in 1995. He is the director of the Education Affair Unit–Deanship of Community Services. Dr. Asiri is a member of the advisory committee for advancing materials, National Technology Plan, King Abdul Aziz City of Science and Technology, Riyadh, Saudi Arabia. He is an editorial board member of the *Journal of Saudi Chemical Society, Journal of King Abdul Aziz University, Pigment and Resin Technology Journal, Organic Chemistry Insights, Libertas Academica, and Recent Patents on Materials Science*. Dr. Asiri holds membership in several national and international societies and professional bodies.

Contents

Preface

The modern world is moving towards sustainable development and furan is a key material in this transition. Furan is processed from furfural, which is an organic compound obtained from biomass feedstock. Thus, furan is a green and environmentally friendly material. It is used to produce pharmaceuticals, resin, agrochemicals, and lacquers. It is an important starting material for a variety of industries for the preparation of many useful products. This book presents comprehensive information on furan and its derivatives.

Chapter 1 presents recent developments in catalytic enantioselective reactions of furans derived from biomass, such as unsubstituted furan, 2-methylfuran, 2,5-dimethylfuran, and furfural. Although several review articles have dealt with the Diels-Alder reactions of furans, there have been no articles highlighting enantioselective versions. The resulting products derived from the catalytic enantioselective reaction of furan are often found as core structures in natural products and pharmaceuticals with important pharmacological activities. After recognizing the valuable skeleton of chiral furan derivatives, numerous attempts have been made to synthesize them by utilizing enantioselective cycloaddition reactions, Friedel-Craft reactions, and nucleophilic addition reactions. Enantioselective cyclization reactions using furans as the 4π diene component provided chiral dihydrofuran derivatives. On the other hand, Friedel-Craft and nucleophilic addition reactions served various furan derivatives with the chiral carbon atom in the a-position.

Chapter 2 examines the synthesis of the compound 2-[2-(4-methylphenylamino)-4-phenylthiazol-5-yl]benzofuran prepared from 1-(4-methylphenyl)-3-(N-phenylbenzimidoyl)thiourea and 2-(2-bromoacetyl)benzofuran in the presence of triethylamine and characterized by FTIR, NMR, and mass spectra. Density functional theory (DFT) computations were adopted for the geometric optimization of this compound to evaluate the Mulliken atomic charge distribution, HOMO-LUMO energy gap, and vibrational analysis. The titled compound induced G1 cell cycle arrest, which is regulated by CDK2 in cancer cells. Therefore, we used molecular modelling to study in silico the possible inhibitory effect as a mechanism of this compound as anticancer agents (PDB code:2KW6, 6DL7, 6VJO, 6WMW, 7LAE). The molecular docking study revealed that the compound was most effective in inhibiting CDk2 cancer cells.

Chapter 3 reports on the synthesis of new α,α-diaminoester and α,α-diamino acid derivatives, as 2-benzamido-2-[(tetrahydro-furan-2-ylmethyl)amino] acetic acid through alkaline hydrolysis reaction of corresponding N-benzoylated methyl α,α-diamino ester. The α,α-diaminoester derivative was synthesized by nucleophilic substitution of methyl α-azido glycinate N-benzoylated with 2-tetrahydrofuran-2-ylmethan-amine. The structure of these products was established on the basis of NMR spectroscopy (^{1}H, ^{13}C) and MS data.

Chapter 4 describes new reactive hotmelt (RHM) adhesives based on thermally reversible Diels-Alder networks comprising multifunctional furan and maleimide prepolymers. The prepolymer mixture is easy to apply in bulk from the melt and after application to the substrates, the adhesive undergoes polymerization at room temperature resulting in crosslinked bonds. Due to their thermoplastic nature and low melt viscosity at hot melt application temperatures, the adhesives provide processing properties similar to moisture-cured polyurethanes (PURs). The technology is isocyanate-free and does not require moisture to initiate the crosslinking. Bonding and tensile properties of the RHM adhesive can be readily tuned by prepolymer design and provide cure rates like those of PUR adhesives. The Diels-Alder adhesives provide versatile adhesion to a variety of substrates and good creep resistance up to the retro temperature. The adhesives show good thermal stability during application and can be recycled multiple times by simple heating/ cooling of the bonds providing similar performance. Several furan and maleimide prepolymers were scaled up to multi-Kg quantities to demonstrate the potential for industrial scalability. The results demonstrate that furan-maleimide reversible chemistry can be used for RHM application as a more sustainable alternative to conventional moisture curing PURs, which tend to contain harmful residual isocyanate monomers.

Chapter 5 discusses the evolution of NOx precursors NH_3 and HCN from pyrolysis of furfural residue (FR). The pyrolysis process was carried out in a thermogravimetric analyzer (TGA) coupled to a Fourier-transform infrared (FTIR) spectrometer. The combination revealed insightful information on the evolution of NH_3 and HCN. This information allows for better understanding of the characteristics of FR derived from furfural production, especially with regard to NH_3 and HCN. Nitrogen is considered a minor component in biomass wastes; in this study nitrogen content is about 0.57%. However, the pollution potential of low-nitrogen content is huge through both direct and indirect processes. Thus, this study presents results on FR pyrolysis in a pure nitrogen environment. At the heating rate of 40°C/min^{-1}, the only NOx precursor detected was HCN at 713cm^{-1} as per the database provided by National Institute of Standards and Technology (NIST). NH_3 was not detected. The particle size of FR used ranged between 0.15 and 0.25 mm.

Chapter 6 reviews pure furanics, tannin–furanics, and tannin–furanic-furanic humins as fire resistant, environmentally friendly rigid biofoams. More recently, furanic wood adhesives have been developed in which a major furan portion is coupled with either synthetic or bioadhesives. In the case of furanic wood bioadhesives, formulations developed range from being 90% to 100% biosourced. Equally, furanic rigid plastics of considerable mechanical resistance have also been developed and applied in the production of angle-grinder disks and automotive brakes with very encouraging results showing the capacity of these resins to resist to the very high mechanical stresses applied.

Chapter 7 describes furfural, a five-membered heterocyclic aromatic hydrocarbon derivable from acid hydrolysis of sugar cane bagasse, maize cob, rice husk, or any cellulose-containing material. It is useful in the synthesis of a range of specialized chemical products. Its condensation with nitromethane in basic medium yields 2-(2-Nitrovinyl) furan. This functional group (nitrovinyl) has been documented as a potent antimicrobial agent against gram-positive and gram-negative bacteria, with more potency against the gram-positive strains. The reaction of urea and thiourea with furfural yields bisimines-1,3-bis[(*E*)-furan2-yl)methylene] urea, and 1,3-bis[(*E*)-furan-2-yl) methylene]thiourea, respectively. The two compounds are good antimicrobial agents in addition to the latter being a potential dye for wool and cotton fabrics with different hues. Also, the reaction between acetophenone and furfural (an aldehyde) in a basic medium yields the chalcone: (*E*)-3-(furan-2-yl)-1-phenylprop-2-ene-1-one. This chalcone has been confirmed as a good antifungal agent and wood-protector against termite attack. Thus, chemical modification of the aldehyde functional group of furfural to nitro, imine, and chalcone groups imparted different activities on furfural.

Anish Khan
Center of Excellence for Advanced Materials Research,
King Abdulaziz University,
Jeddah, Saudi Arabia

Mohammed Muzibur Rahman
and Abdullah Mohammed Ahmed Asiri
King Abdulaziz University,
Jeddah, Saudi Arabia

Salman Ahmad Khan
School of Sciences,
Maulana Azad National Urdu University,
Hyderabad, Telangana, India

M. Ramesh
Kalaignarkarunanidhi Institute of Technology,
Kannampalayam, India

Section 1

Furans and Furan Derivatives - Recent Advances

Chapter 1

Catalytic Enantioselective Reactions of Biomass-Derived Furans

Dong Guk Nam, Jung Woon Yang and Do Hyun Ryu

Abstract

In this chapter, recent developments with regard to catalytic enantioselective reactions of furans, derived from biomass such as unsubstituted furan, 2-methylfuran, 2,5-dimethylfuran, and furfural are described. Although several review articles have dealt with the Diels-Alder reactions of furans, there have been no articles highlighting enantioselective versions. The resulting products derived from the catalytic enantioselective reaction of furan are often found as core structures in natural products and pharmaceuticals with important pharmacological activities. After recognizing the valuable skeleton of chiral furan derivatives, numerous attempts have been made to synthesize them by utilizing enantioselective cycloaddition reactions, Friedel-Crafts reactions, and nucleophilic addition reactions. Enantioselective cyclization reactions using furans as the 4π diene component provided chiral dihydrofuran derivatives. On the other hand, Friedel-Crafts and nucleophilic addition reactions served various furan derivatives with a chiral carbon atom in the α-position.

Keywords: enantioselective, cycloaddition, Diels-Alder, Friedel-Crafts, furan, nucleophilic addition

1. Introduction

Furfural and 5-hydroxymethylfurfural (HMF) have received significant attention as promising platform chemicals due to their versatile utility in the synthesis of various commodity chemicals and fuels [1–3]. These platform chemicals can be easily transformed into value-added chemicals, such as 2-methylfuran, 2,5-dimethylfuran, and other furans via chemical conversions or fermentation [4–6]. Since aromatic heterocycle furans are present in a variety of chiral natural products, pharmaceuticals, and other intermediates, a plethora of enantioselective methodologies has been developed for the synthetic community [7–9]. The important strategies are given as follows—(i) enantioselective cyclization reactions including cycloadditions using furans as the 4π diene component and cyclopropanation between furan and diazoester to obtain various valuable chiral synthons (Section 2); (ii) enantioselective Friedel-Crafts cycloadditions for the fabrication of carbon-carbon bonds between furans and electron-deficient alkenes, yielding chiral centers at the α- or β-position of furans (Section 3); (iii) various enantioselective nucleophilic addition reactions of furfural as an electrophile for the construction of chiral hydroxyl functional groups (Section 4). Thus, this chapter is divided into three sections.

2. Catalytic asymmetric cyclization reactions of furans

Since the first cyclopropanation between unsubstituted furan and chlorodiazopropene was reported by de Meujere and Kositkov in 1991, reactions using unfunctionalized furan have emerged as a challenging area in organic chemistry [10]. In most cases, numerous reports have utilized substituted furan at the 2- or 3-positions. However, biomass-derived furan such as normal furan or methyl-substituted furans are generally held to be poor dienes in Diels-Alder reactions and have poor reactivity for cyclization as well as cyclopropanation. Therefore, it has been difficult to develop such reactions with simple furan, and extending it to the catalytic enantioselective version was extremely difficult. Since the discovery of an enantioselective furan Diels-Alder reaction in 1997 by the Evans group [11], some progress in this area has been achieved. The aim of this chapter is to mainly discuss the catalytic enantioselective reaction of simple furans for Diels-Alder reaction, [4 + 3] cyclization, and cyclopropanation. The reaction of functionalized and substituted furan will not be included here.

2.1 Cu or Pd-catalyzed enantioselective Diels-Alder reactions with furans

The first highly enantioselective catalytic Diels-Alder reaction using an unsubstituted furan reactant was accomplished by the Evans group in 1997 [12, 13]. They utilized a bisoxazoline-copper complex **1** as a Lewis acid catalyst for the Diels-Alder reaction between acrylamide **3** and furan **2** to produce the chiral cycloadduct products **4** as an important synthetic intermediate of shikimic acid in 97% yield with 97% ee (**Figure 1**).

The Diels-Alder reaction between acrylamide **3** and furan **2** was accomplished using different metal catalysts. The Lassaletta and Ishihara groups independently reported copper(II) complex-catalyzed Diels-Alder reactions to produce the *endo*-selective cycloadduct product **8** in 92% yield with 98% ee and 88% yield with 96% ee, respectively [14, 15]. An *exo*-selective and highly enantioselective Diels-Alder reaction of acrylamide **3** and furan **2** were accomplished by the Kabuto group in 2004 through the use of a chiral phosphinooxazolidine-palladium complex **7** as the active catalyst (**Figure 2**) [16].

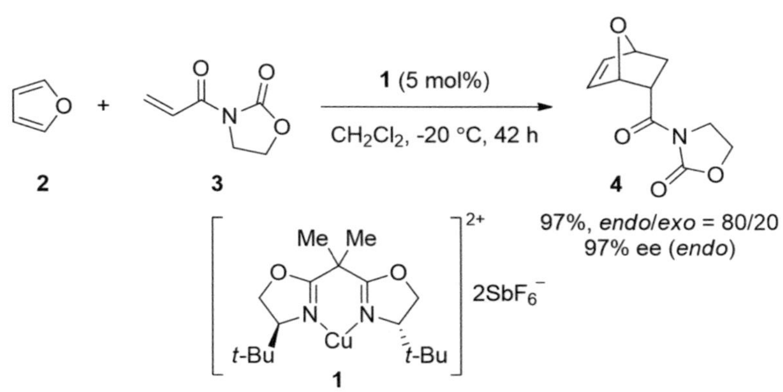

Figure 1.
Catalytic enantioselective Diels-Alder reactions with copper catalyst.

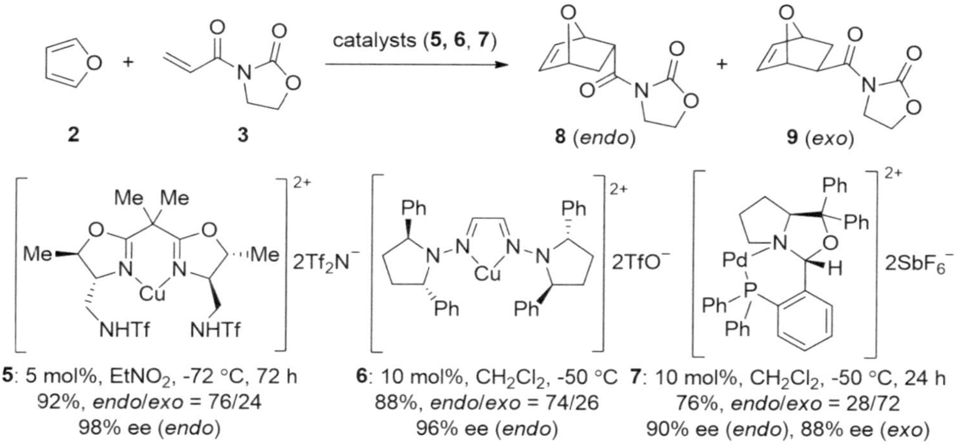

Figure 2.
Chiral copper or palladium catalyst for the catalyzed Diels-Alder reaction.

2.2 Oxazaborolidium-catalyzed enantioselective Diels-Alder reactions with furans

A few examples of catalytic asymmetric Diels-Alder reaction of unsubstituted furans have been reported. Corey, Ryu, and coworkers successfully reported the Diels-Alder reaction between furan **12** and 1,1,1-trifluoroethyl acrylate **3** through the use of oxazaborolidinium **10** or **11** as a metal-free catalyst in combination with trifluoroacetic acid (TFA) or bis(trifluoromethane)sulfonimide (Tf_2NH) (**Figure 3**) [17]. Various methyl-substituted furans were employed as dienes, which exhibited superior activity and tolerance for this study, rendering the desired cycloadduct product **13** in excellent yields with excellent diastereo- and enantioselectivities.

In 2011, Shibatomi and coworkers accomplished chiral oxazaborolidine **15**-catalyzed enantioselective Diels-Alder reactions between furan **12** and fluoromethylated (*E*)- or (*Z*)-acrylate yielding the corresponding product **17** with up to 99% ee (**Figure 4**) [18]. As depicted in **Figure 4**, various β-fluoro-substituted

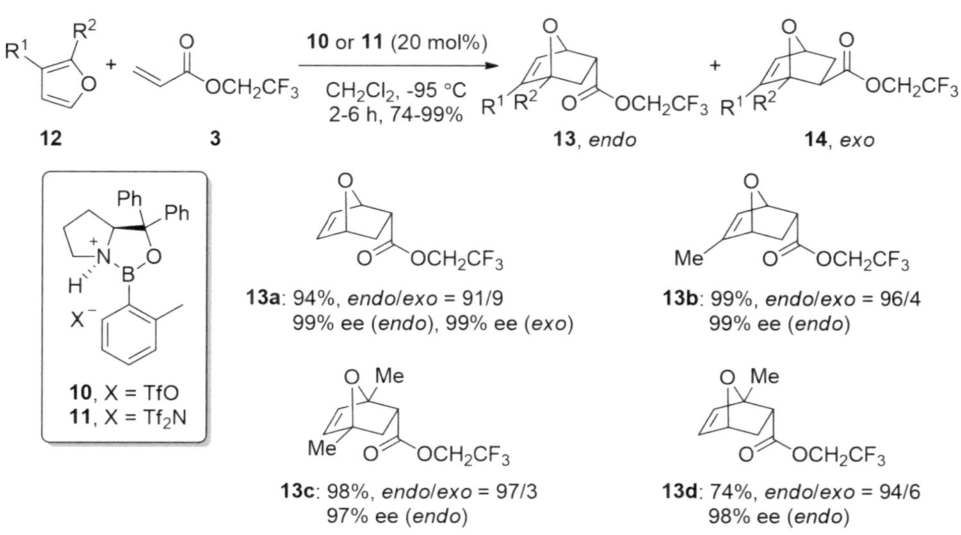

Figure 3.
Enantioselective Diels-Alder reactions with 1,1,1-trifluoroethyl acrylate.

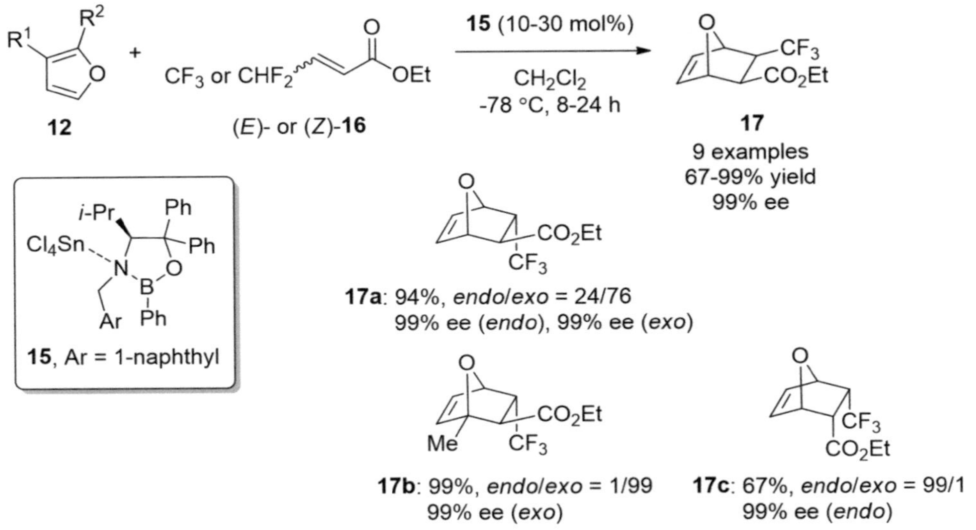

Figure 4.
exo-*Selective enantioselective Diels-Alder reaction with fluoromethylated acrylate.*

(*E*)- or (*Z*)-acrylates **16** and substituted furans **12** were well tolerant with a selective approach for high enantioselectivities as well as *endo*/*exo*-selectivities (up to 99/1 and 1/99).

Corey and coworkers reported asymmetric Diels-Alder reactions of di-substituted furans **20** with acrylate **21** in 2016 [19]. The use of oxazaborolidinium catalyst **18** activated by aluminum bromide ($AlBr_3$) gave the cycloadduct **22** in 99% yield with 99% ee (**Figure 5**). Diastereoselectivity and reaction times were further improved through the introduction of fluorinated oxazaborolidines as second-generation catalyst **19**.

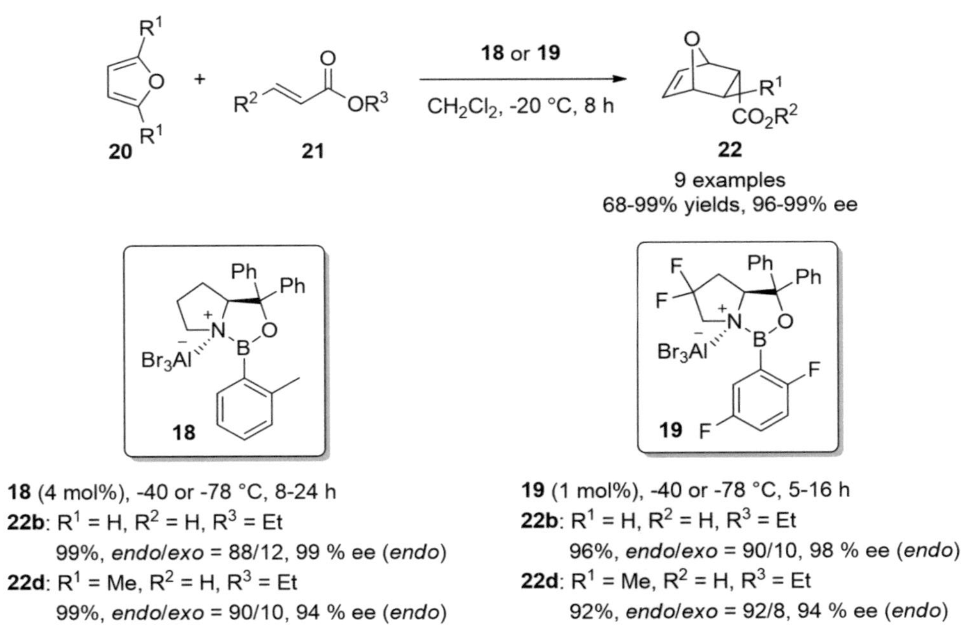

Figure 5.
Enantioselective Diels-Alder reactions with various dienophiles.

Occasionally, the catalytic system comprising a chiral *N*-heterocyclic stabilized borenium cation for the enantioselective Diels-Alder reaction required low reaction temperatures. To overcome this drawback, Chein and coworkers designed a sulfur-stabilized borenium cation, oxathiaborolium catalyst **23** in combination with tin chloride ($SnCl_4$). However, in the case of unsubstituted furan **2**, the reaction required −60 °C for the enantioselective Diels-Alder reaction with ethyl acrylate **24** (**Figure 6**) [20].

In 2010, Corey and coworkers reported a catalytic asymmetric Diels-Alder reaction by employing an allenic ester **26** as the dienophile with di-substituted furans **20**. The use of 5–20 mol% of chiral oxazaborolidinium ion (COBI) **11** or **18** as a catalyst gave various synthetically valuable cycloadducts **27** with good to excellent yields and high stereoselectivities (**Figure 7**) [21].

The usefulness of the Diels-Alder cycloadduct **27a** is illustrated in **Figure 8**. Selective reduction of **27a** and hydrogenation using Wilkinson's catalyst produced synthetic unit **28**. Further transformation of **28** to (−)-laurenditerpenol, known to be a potent inhibitor of HIF-1α, was achieved based on a known procedure [22].

An alternative organocatalytic Diels-Alder reaction of furan **2** with acrylic enone **29** was developed by the Harada group. *Allo*-Threonine-derived oxazaborolidinones (OXB) **28** were employed as a catalyst to afford the corresponding cycloadduct **30** with good to high yields and excellent enantioselectivities (**Figure 9**) [23]. Although this new motif catalyst **28** has weaker Lewis acidity compared to the cationic

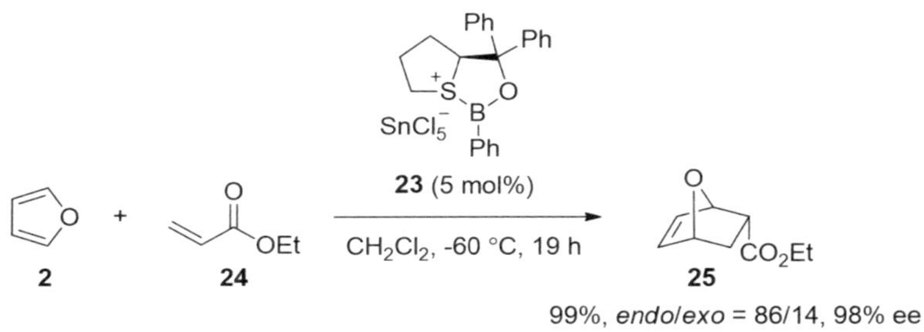

Figure 6.
Enantioselective Diels-Alder reactions with ethyl acrylate.

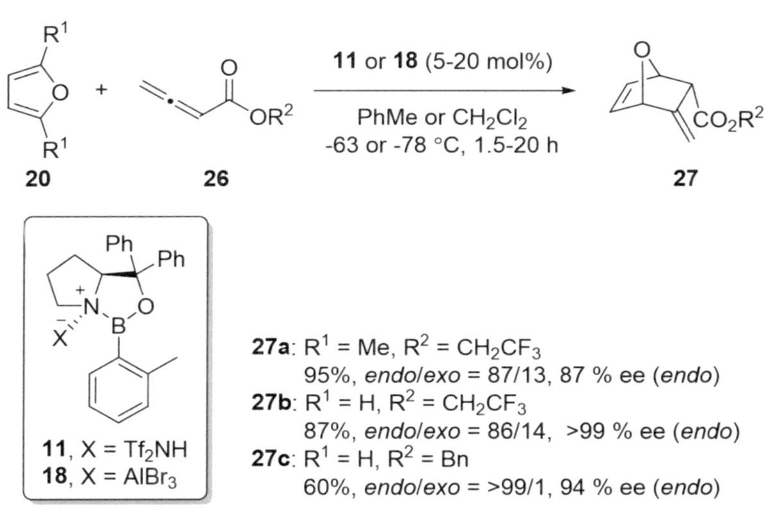

Figure 7.
Enantioselective Diels-Alder reactions with allenic ester.

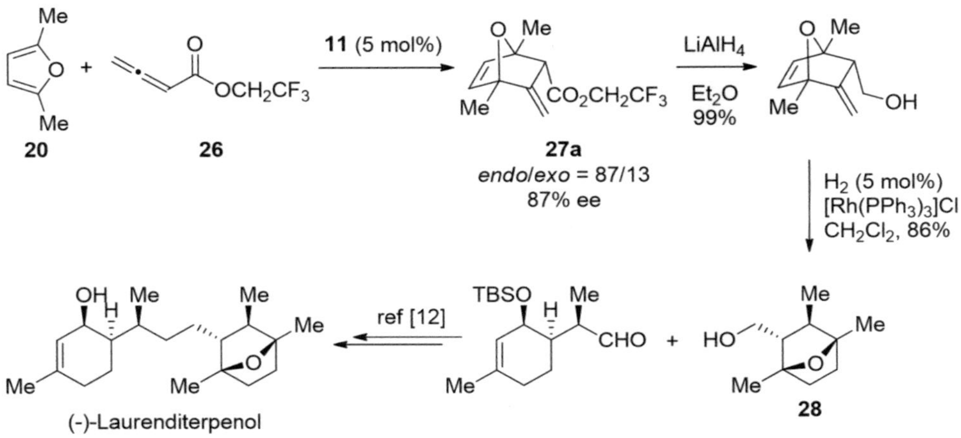

Figure 8.
Synthesis of (−)-lauenditerpenol.

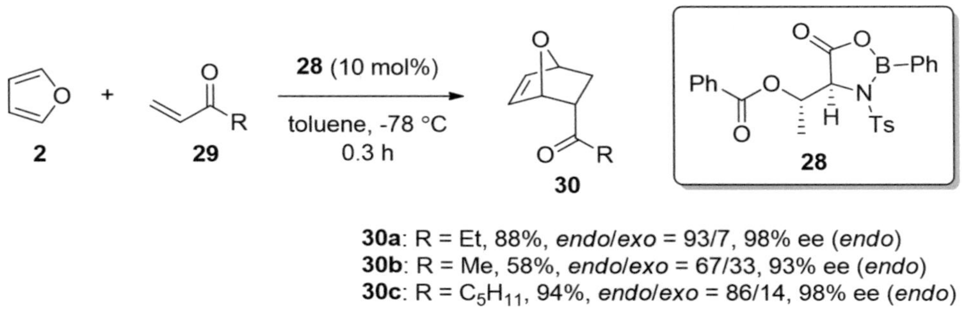

Figure 9.
Chiral Diels-Alder reactions with α,β-unsaturated ketones catalyzed by oxazoborolidinone.

oxazaborolidine catalyst, OXB catalyst **28** exhibited high performance in terms of stereoselectivity in Diels-Alder reactions between furans and α,β-unsaturated ketones.

2.3 Enantioselective [4 + 3] cyclization (or annulation) reactions with furans

The [4 + 3]-annulation consisting of the tandem cyclopropanation/Cope rearrangement of furan is a useful and predictable tool for the stereoselective synthesis of seven-membered rings. Asymmetric synthesis of 8-oxabicyclo[3.2.1] octene derivatives (**33** or **34**) was achieved by utilizing vinyl diazoacetate **31** or **32** bearing chiral auxiliaries, such as (*S*)-lactate or (*R*)-pantolactone, respectively, in the presence of catalytic amounts of rhodium(II) octanoate. Practical and general [3 + 4]-annulation methods for the synthesis of oxabicyclic product with excellent yields (up to 91% yield) and enantioselectivities (up to 95% ee) were developed by Davies and coworkers in 1996 (**Figure 10**) [24].

In 2008, the same group described the Tetrakis[(*R*)-(+)-*N*-(*p*-dodecylphenylsulfonyl)prolinato]dirhodium(II) (Rh_2(*R*-DOSP)$_4$)-catalyzed reaction of vinyl diazoacetate **35** and furan **36** for the generation of formal [4 + 3] cycloadducts **37** with excellent stereoselectivities (up to >94% de and 98% ee). This reaction was smoothly proceeded by a tandem cyclopropanation/Cope rearrangement followed by stereoselective tautomerization (**Figure 11**) [25].

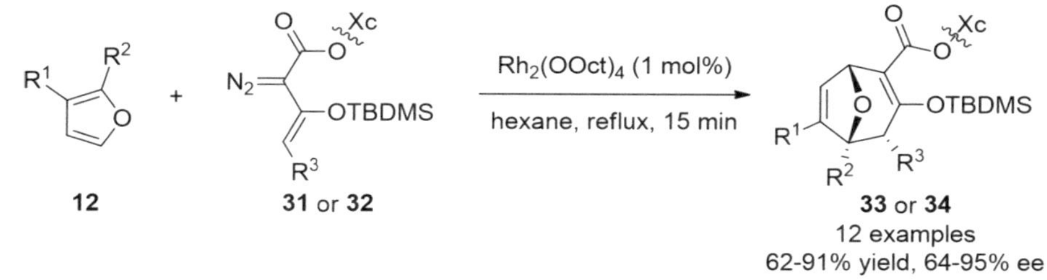

31: Xc = (*S*)-lactate
33a: R^1 = H, R^2 = H, R^3 = H, 72%, 79% ee
33b: R^1 = H, R^2 = H, R^3 = Me, 62%, 90% ee
33c: R^1 = H, R^2 = Me, R^3 = Me, 91%, 84% ee

32: Xc = (*R*)-pantolactate
34a: R^1 = H, R^2 = H, R^3 = H, 82%, 94% ee
34b: R^1 = H, R^2 = H, R^3 = Me, 75%, 95% ee
34c: R^1 = H, R^2 = Me, R^3 = Me, 69%, 94% ee

Figure 10.
Enantioselective [4 + 3] cyclization with chiral auxiliary substituted diazoacetoacetate.

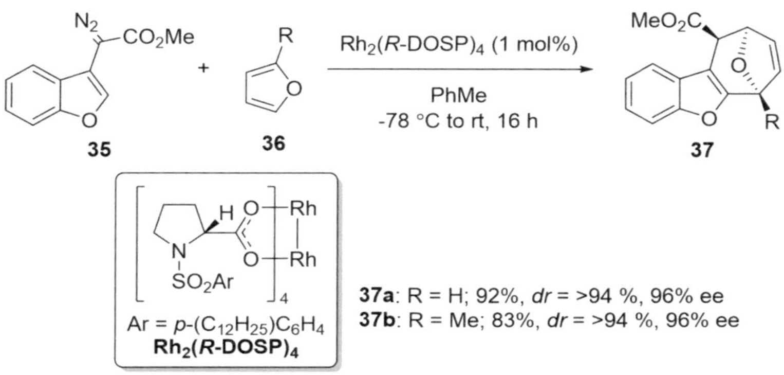

Figure 11.
Enantioselective [4 + 3] cycloaddition with benzofuranyldiazoacetates.

In 2017, Vicario and coworkers reported that chiral 1,1-binaphthol (BINOL)-based Brønsted acid **38** catalyzed the enantioselective oxidative [4 + 3] cycloaddition of furan **40** and oxyallyl cation generated *in situ* through the oxidation of allenamide **39** with dimethyldioxirane (DMDO) as the oxidant. Stereochemical environments were induced through hydrogen-bonding and ion-pairing interactions during the [4 + 3] cycloaddition process, enabling efficient chirality transfer that furnished [4 + 3] cycloaddition products **41** in excellent yields and with high stereocontrol (**Figure 12**) [25].

In 2017, Jacobsen and coworkers reported that H-bond donors such as chiral squaramide **42** could activate relatively unreactive electrophiles for promoting enantioselective reactions in the following manner. Initially, chiral squaramide was able to interact with silyl triflates by binding the triflate counterion to produce a highly Lewis acidic complex (so-called enhanced Lewis acidity). The silyl triflate-chiral squaramide combination promoted the generation of oxocarbenium intermediates from acetal **43**. Controlled enantioselectivity during the nucleophilic addition of furan **40** to the cationic intermediate was achieved through noncovalent interactions between the squaramide catalyst and the oxocarbenium triflate. Under optimal reaction conditions, the cycloadducts **44** could be obtained in 55–98% yields with 66–96% ee (**Figure 13**) [26].

2.4 Enantioselective cyclopropanation reactions with furans

Reactions of furans with carbenoids led to cyclized reactions, such as cyclopropanation. Additionally, a cyclopropanation reaction could be performed through

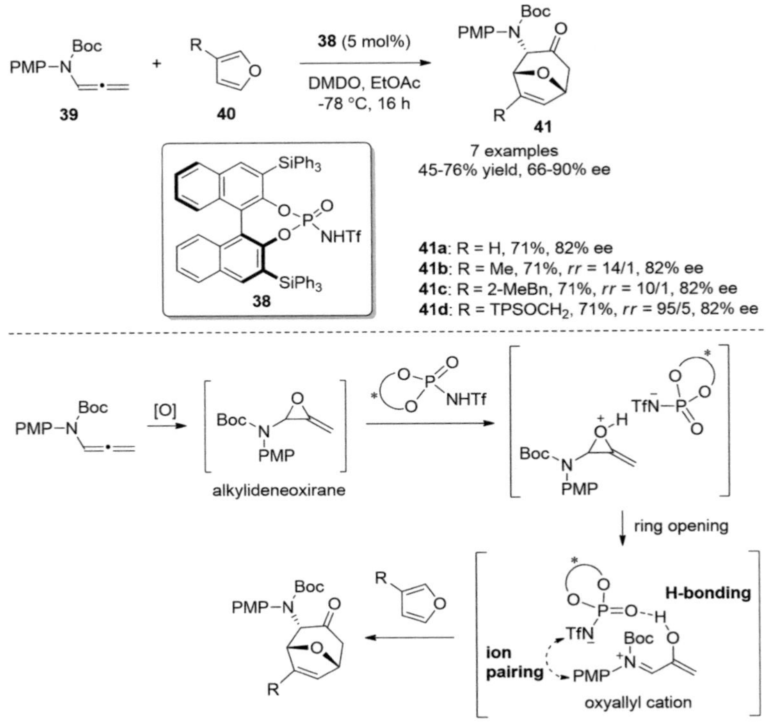

Figure 12.
Enantioselective [4 + 3] cyclization with allenamide.

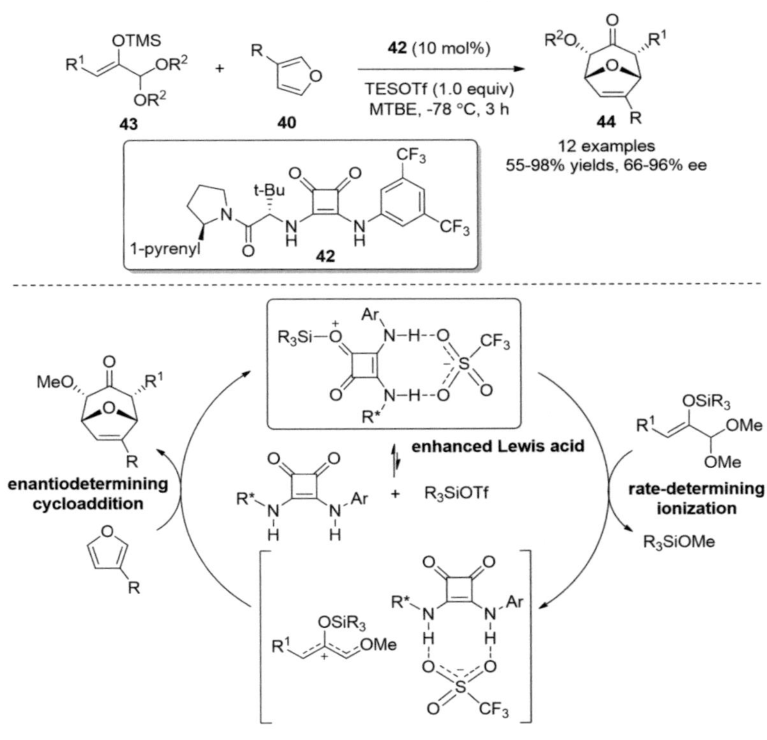

Figure 13.
Enantioselective [4 + 3] cyclization with silyl enol ether.

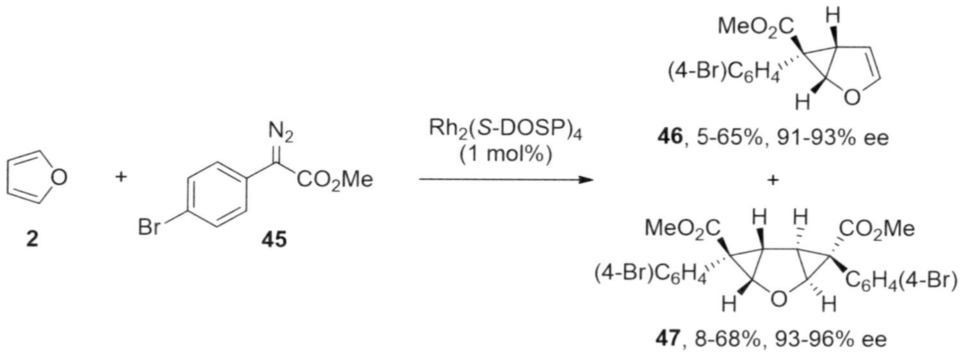

Figure 14.
Enantioselective cyclopropanation with vinyldiazoacetates.

the reaction of furan and diazoacetate under a metal catalyst. Reiser and coworkers reported the enantioselective cyclopropanation of furans using a copper catalyst, however, the reaction was achieved when the furan was substituted with ester groups at the 2- or 3-position [27]. To solve this problem, Davies and coworkers designed the catalytic system using dirhodium catalyst. When simple furan and aryl diazoester was subjected to the rhodium-catalyzed enantioselective cyclopropanation reaction, both cycloadduct product **46** and bis-cyclopropanation product **47** were obtained in 5–65% yields with 91–93% ee and 8–68% yields, 93–96% ee, respectively (**Figure 14**) [28].

3. Catalytic asymmetric Friedel-Crafts reactions of furans

One of the most efficient methods for the synthesis of chiral heteroaromatic compounds with a stereogenic center in the benzylic position is the Friedel-Crafts reaction between carbonyl compounds and electron-deficient alkenes [29]. This field of chemistry has been intensely explored since around 2000, and interest in this field is still growing. Most catalytic enantioselective Friedel-Crafts reactions can be utilized with electron-rich aromatic and heteroaromatic compounds, such as aniline and indole derivatives. However, reports with regard to the use of furans for this study are still scarce due to the relative instability and reduced nucleophilicity of furans compared to indoles and pyrroles [30]. In particular, catalytic enantioselective versions of the Friedel-Crafts reaction with biomass-derived furans as well as normal furan are much less developed than other aromatics.

The first catalytic enantioselective Friedel-Crafts reaction using biomass-derived furan was accomplished by the Jørgensen group in 2000 [31]. Only methyl or trimethylsilyl-substituted furans **36** were subjected to the Friedel-Crafts reaction in combination with ethyl glyoxalate **49** in the presence of the C2-symmetric chiral Cu(II)-bis(oxazoline) complexes **48** as the catalyst, resulting in the formation of the desired product **50** in low to high yields with moderated enantioselectivities (**Figure 15**).

One year later, in 2001, the same group described an enantioselective Friedel-Crafts reaction of normal furan or biomass-derived furans **36** with ethyl trifluoropyruvate **51** utilizing the chiral Cu(II)-bis(oxazoline) complex **48** [32]. In the case of non-substituted furan, a poor yield (15%) for the Friedel-Crafts product **52** was observed despite achieving good enantioselectivity (81% ee). However, various substituted furans provided the desired products with good to high enantioselectivities (**Figure 16**).

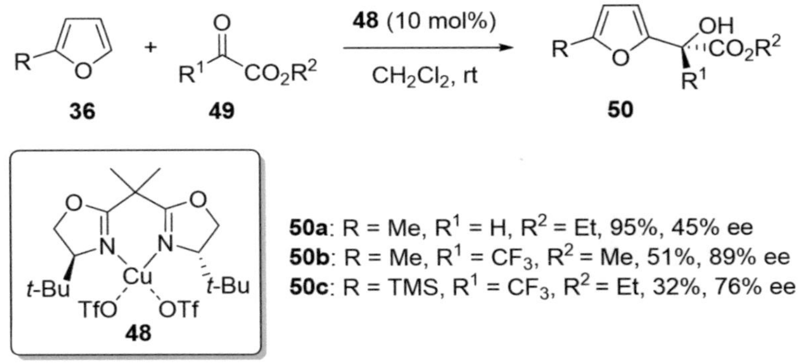

Figure 15.
Enantioselective Friedel-Crafts reactions with glyoxalates.

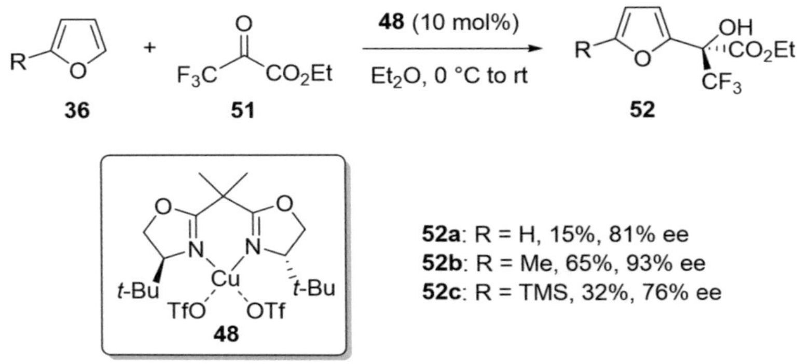

Figure 16.
Enantioselective Friedel-Crafts reactions with ethyl trifluoropyruvate.

In 2009, Yamazaki and coworkers utilized chiral *cis*-aminoindanol-derived bis (oxazoline)-Cu(II) complexes in catalytic enantioselective Friedel-Crafts reactions between furans **12** and ethenetricarboxylates **54**. As a result, chiral 2-alkylated products **55** were obtained in high yields (73–93) with low to moderate enantioselectivities (25–62% ee) (**Figure 17**) [33].

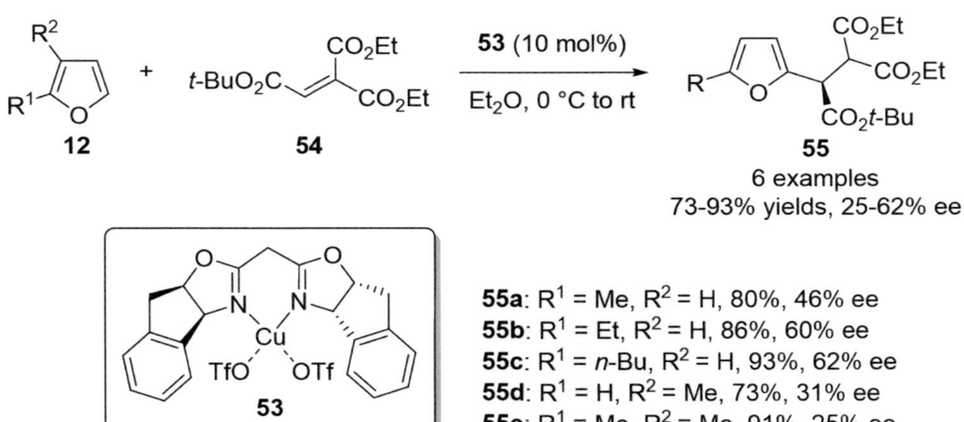

Figure 17.
Enantioselective Friedel-Crafts reactions with ethenetricarboxylates.

Another attempt with regard to the asymmetric Friedel-Crafts reaction of furans **36** with alkyl glyoxalates **57** utilizing Jacobsen's Co(II)-salen complexes **56** as a catalyst was accomplished by the Jurczak group in 2006 (**Figure 18**) [34]. High-pressure (ca. 10 kbar) conditions were essential to obtain chiral furfuryl alcohols **58** as an important synthetic intermediate in moderate to good yields (28–85%) with moderate enantioselectivities (26–66% ee).

A few years later, in 2008, the same group successfully performed catalytic enantioselective Friedel-Crafts reaction between furans **36** and *n*-butyl glyoxalates **60** by switching the catalytic system from Co(II)-salen complexes **56** to BINOL/Ti complexes **59**. As a result, the enantioselectivity and chemical yield of the desired chiral furanyl hydroxyacetate **61** were enhanced compared to the previous results in **Figure 18**. Notably, various substituted furans including normal furan **36** were tolerant for this reaction and provided the desired products **61** in excellent yields with high to excellent enantioselectivities (**Figure 19**) [35].

Cationic square planar metal complexes [M(diphosphine)]$^{2+}$, where M = Pt, Pd, Ni)] have emerged as an alternative class of Lewis acid catalysts such as

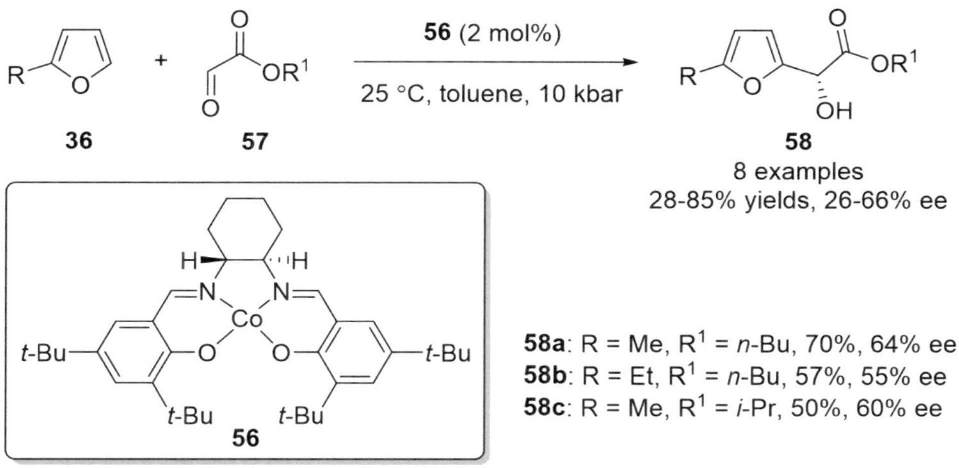

Figure 18.
Enantioselective Friedel-Crafts reactions with alkyl glyoxalates.

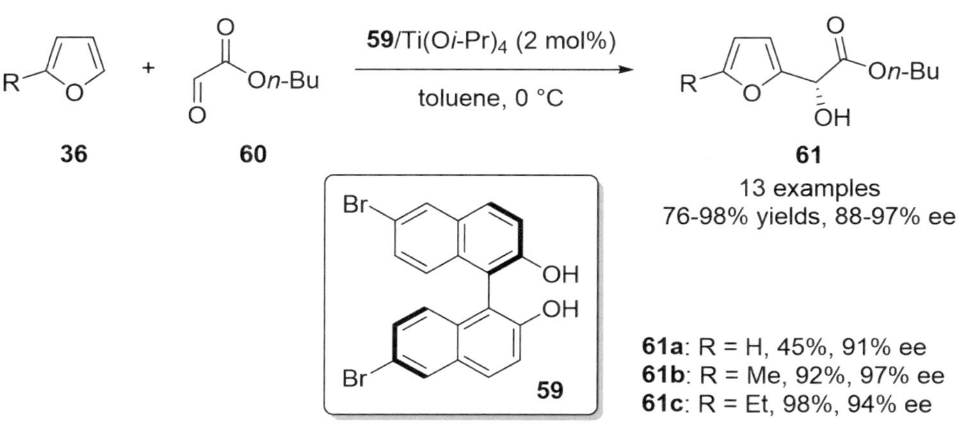

Figure 19.
Enantioselective Friedel-Crafts reactions with n-*butyl glyoxalates.*

Cu-bisoxazolines (Box), Ti-BINOL, and Co-salen due to the following unique characteristics—(i) well-defined coordination geometries to help control the stereochemical environment; (ii) high carbophilicity; and (iii) tunable electronic properties for enhancing Lewis acidity [36]. Mehdi-Zodeh and coworkers introduced cationic square planar-platinum or palladium metal complexes as Lewis acid catalysts into the Friedel-Crafts reaction between biomass-derived furans **36** and ethyl trifluoropyruvate **51**. Specifically, the use of either 2,2′-bis (diphenylphosphino)-1,1′-binaphthyl (BINAP) **62** or 9,9′-dimethyl-9,9′,10,10′-tetrahydro-9,10,9′,10′-biethenobianthracene-11,11′-bis(diphenylphosphino)-12,12′-diyl (Me_2-CATPHOS) **62** gave the corresponding 2-hydroxy-trifluoromethyl ethyl esters **64** in good yields with moderate to good enantioselectivities (**Figure 20**) [37].

The first enantioselective organocatalytic Friedel-Crafts reaction with biomass-derived furan **66** using the first-generation MacMillan's chiral imidazolidinone as an organocatalyst **65** was reported by the thesis of Paras in 2004 [38]. In general, the sense of high asymmetric induction using a chiral imidazolidinone catalyst for enantioselective reactions was well established with the following distinctive features—(i) *E*-selective iminium ion formation when reacting the catalyst with α,β-unsaturated aldehydes; (ii) chirality of the benzyl group on the catalyst backbone shields *re*-face of the α,β-unsaturated iminium ion, leaving the *si*-face exposed to nucleophilic addition. However, the desired Friedel-Crafts product **68** was unfortunately obtained in high yield but moderate enantioselectivity when employing biomass-derived furan **66** (**Figure 21**).

In 2010, Harada and coworker reported an organocatalytic Friedel-Crafts reaction between furans **12** and α,β-unsaturated ketones **70** using a chiral oxazaborolidinone (OXB) catalyst **69** to produce the chiral Friedel-Crafts products **71** in good to excellent yields (62–99%) with high enantioselectivities (77–93% ee) [39]. As shown in **Figure 22**, different substituted furans and α,β-unsaturated ketones were well tolerated in this reaction.

The highly enantioselective organocatalytic Friedel-Crafts reaction with biomass-derived furan **66** using chiral phosphoric acid **72** as an organocatalyst was accomplished by the Akiyama group in 2010 [40]. They utilized a highly sterically hindered phosphoric acid catalyst **72** in the Friedel-Crafts reaction of furan **66** with methyl trifluoropyruvate **73** to afford the desired product **74** in excellent yield of 99% with high enantioselectivity (82% ee) (**Figure 23**).

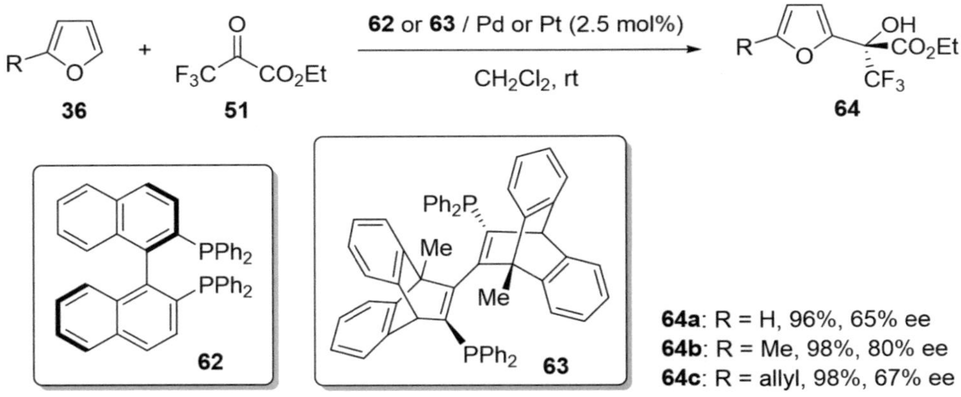

Figure 20.
Enantioselective Friedel-Crafts reactions catalyzed by metal complexes catalyst.

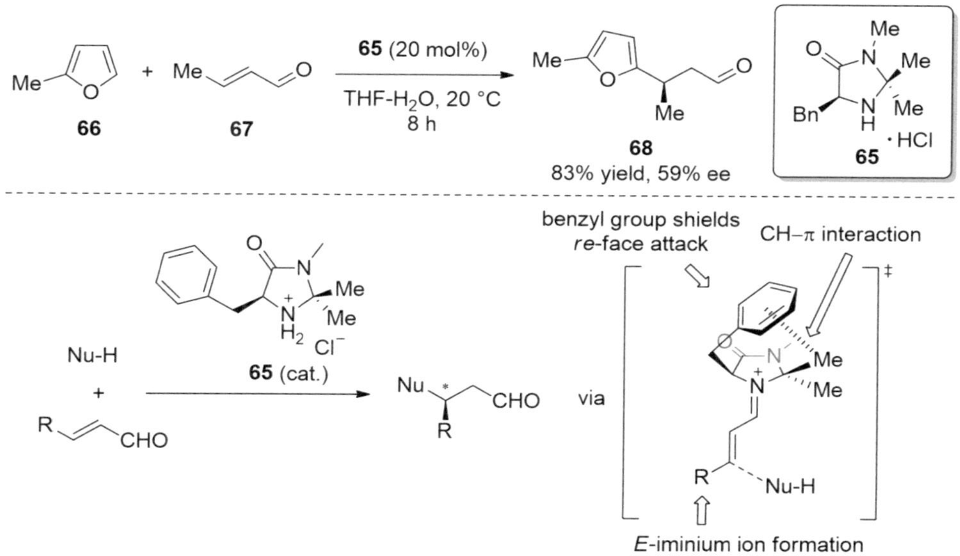

Figure 21.
Enantioselective Friedel-Crafts reactions with α,β-unsaturated aldehydes.

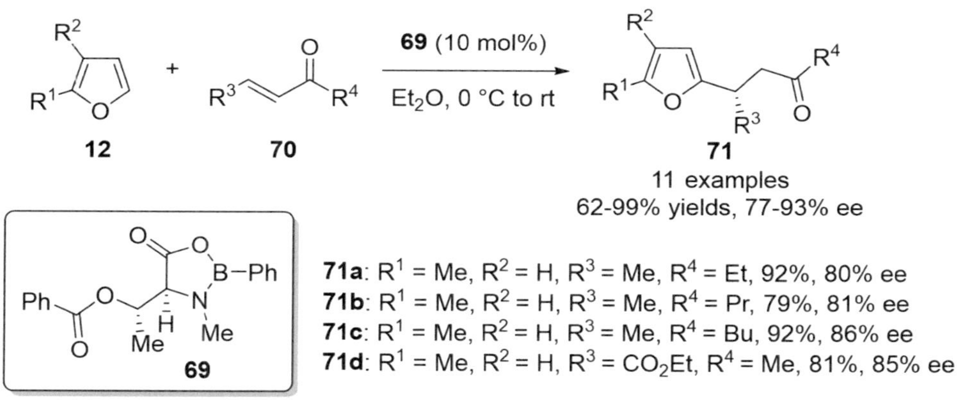

Figure 22.
Enantioselective Friedel-Crafts reactions with α,β-unsaturated ketones.

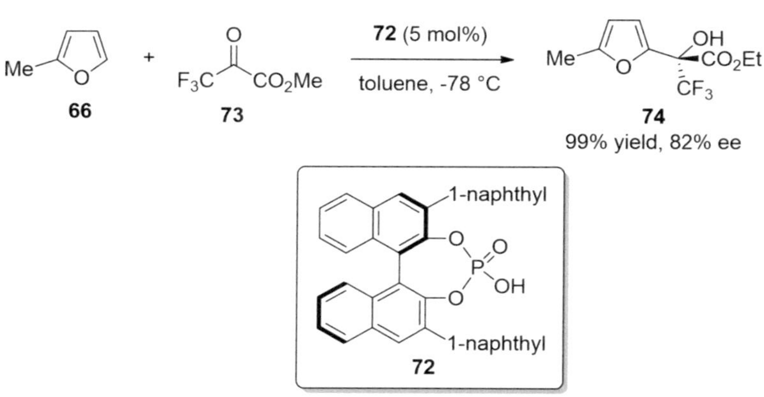

Figure 23.
Enantioselective Friedel-Crafts reactions catalyzed by phosphoric acid.

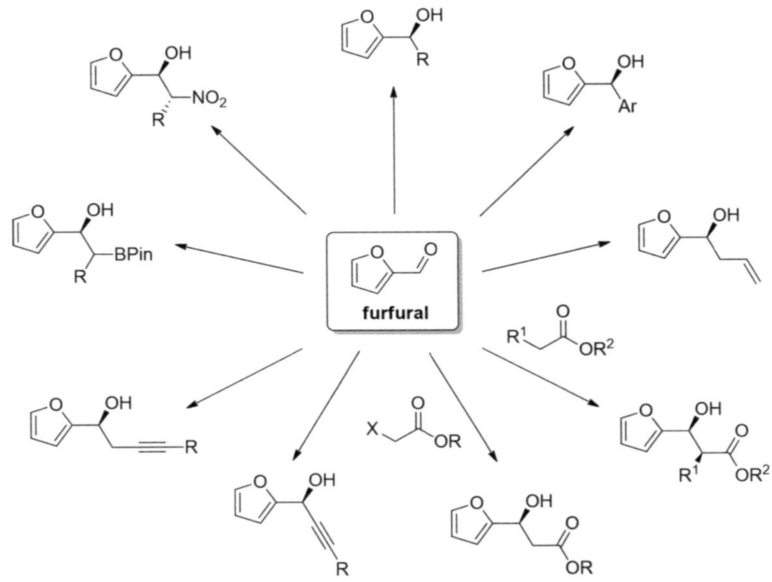

Figure 24.
Various catalytic asymmetric nucleophilic addition reactions with furfural.

4. Various catalytic asymmetric nucleophilic addition reactions of furfural

Asymmetric nucleophilic addition reactions with aromatic or heteroaromatic aldehyde derivatives are powerful C-C bond-forming reactions that can provide chiral hydroxy compounds with stereogenic hydroxy functional groups. Therefore, the development of asymmetric nucleophilic addition is an ongoing challenge in organic synthesis. Following the first demonstration of the catalytic asymmetric nucleophilic addition with biomass-derived furfural by the Yamamoto group in 1997 [41], numerous reports with regard to catalytic asymmetric reactions of furfural have been published including the reaction of allylation [42], aldol reactions [43, 44], nitroaldol (henry) reaction [45, 46], alkylation [47–49], acylation [50], the Reformatsky reaction [51], the Nozaki-Hiyama reaction [52], alkynylation [53], and hydroboration [54] with various types of catalysts (**Figure 24**). However, the enantioselective catalytic nucleophilic addition reaction of 5-hydroxymethyl-furfural (HMF) has not yet been reported.

5. Conclusion

As we have shown in this book chapter, a variety of synthetic approaches, such as cycloaddition reactions, Friedel-Crafts reactions, and nucleophilic addition reactions, are elegant methodologies that have been efficiently used for the enantioselective reaction of biomass-derived furans. While Friedel-Crafts and nucleophilic addition reactions serve various furan derivatives with a chiral carbon atom in the α-position, enantioselective cyclization reactions using furans as the 4π diene component affords chiral dihydrofuran or tetrahydrofuran derivatives. Synthesizing chiral synthons or highly functionalized products derived from furan may show great potential not only for the creation of new libraries that could lead to the development of biologically active compounds but also for stimulating further research toward versatile applications of these molecules via another asymmetric catalysis. There is no doubt that the further development of catalytic enantioselective reactions with biomass-derived furans will continue to provide exciting results in near future.

Author details

Dong Guk Nam[1], Jung Woon Yang[2*] and Do Hyun Ryu[1*]

1 Department of Chemistry, Sungkyunkwan University, Suwon, South Korea

2 Department of Energy Science, Sungkyunkwan University, Suwon, South Korea

*Address all correspondence to: dhryu@skku.edu; jwyang@skku.edu

References

[1] Bozell JJ, Petersen GR. Technology development for the production of biobased products from biorefinery carbohydrates—The US Department of Energy's "Top 10" revisited. Green Chemistry. 2010;**12**:539-554. DOI: 10.1039/b922014c

[2] Ravasco JMJM, Gomes RFA. Recent advances on Diels-Alder-Driven preparation of bio-based aromatics. Chemistry-Sustainability-Energy-Materials. 2021;**14**(15):3047-3053. DOI: 10.1002/cssc.202100813

[3] Hu D, Zahng M, Xu H, Wang Y, Yan K. Recent advance on the catalytic system for efficient production of biomass-derived 5-hydroxymethylfurfural. Renewable and Sustainable Energy Reviews. 2021; **147**:111253. DOI: 10.1016/j.rser.2021.111253

[4] Nakagawa Y, Tamura M, Tomishige K. Catalytic reduction of biomass-derived furanic compounds with hydrogen. ACS Catalysis. 2013;**3**: 2655-2668. DOI: 10.1021/cs400616p

[5] Fang W, Riisager A. Recent advances in heterogeneous catalytic transfer hydrogenation/hydrogenolysis for valorization of biomass-derived furanic compounds. Green Chemistry. 2021;**23**: 670-688. DOI: 10.1039/d0gc03931d

[6] Zakrzewska ME, Bogel-Lukasik E, Bogel-Lukasik R. Ionic liquid-mediated formation of 5-hydroxymethylfurfurals—A promising biomass-derived building block. Chemical Reviews. 2011;**111**: 397-417. DOI: 10.1021/cr100171a

[7] Wong HNC, Yu P, Yick C-Y. The use of furans in natural product syntheses. Pure and Applied Chemistry. 1999; **71**(6):1041-1044. DOI: 10.1351/pac199971061041

[8] Boto A, Alvarez L. Furan and Its Derivatives in Hetrocycles in Natural Product Synthesis. Weinheim: Wiley-VCH; 2011. pp. 97-152. DOI: 10.1002/9783527634880.ch4

[9] Chen VY, Kwon O. Unified approach to furan natural products via phosphine-palladium catalysis. Angewandte Chemie International Edition. 2021; **60**(16):8874-8881. DOI: 10.1002/anie.202015232

[10] de Meijere A, Schulz T-J, Kostikov RR, Graupner F, Murr T, Bielfeldt T. Dirhodium(II) tetraacetate catalyzed (chlorovinyl)cycloprapantion of enol ethers and dienol ethers—A route to donot-substituted vinylcyclopropanes, ethynylcyclopropanes and cycloheptadienes. Synthesis. 1991;7: 547-560. DOI: 10.1055/s-1991-26514

[11] Evans DA, Barnes DM. Cationic bis (oxazoline)Cu(II) Lewis acid catalyst. Enantioselective furan Diels-Alder reaction in the synthesis of *ent*-Shikimic acid. Tetrahedron Letters. 1997;**38**:57-58. DOI: 10.1016/S0040-4039(96)02259-9

[12] Evans DA, Barnes DM, Johnson S, Lectka T, von Matt P, Miller SJ, et al. Bis (oxazoline) and bis(oxazolinyl)pyridine copper complexes as enantioselective Diels-Alder catalysts: Reaction scope and synthetic applications. Journal of the American Chemical Society. 1999; **121**:7582-7594. DOI: 10.1021/Ja991191C

[13] Sakakura A, Kondo R, Matsumura Y, Akakura A, Ishihara K. Rational design of highly effective asymmetric Diels-Alder catalysts bearing 4,4′-sulfonamidomethyl groups. Journal of the American Chemical Society. 2009;**131**:17762-17764. DOI: 10.1021/ja906098b

[14] Lassaletta JM, Alcarazo M, Ferández R. Glyoxal bis-hydrazones: A new family of nitrogen ligands for asymmetric catalysis. Chemical Communications. 2004;**3**:298-299. DOI: 10.1039/b314249c

[15] Nakano H, Takahashi K, Okuyama Y, Senoo C, Tsugawa N, Suzuki Y, et al. Chiral phosphinooxazolidine ligands for palladium- and platinum-catalyzed asymmetric Diels-Alder reactions. Journal of Organic Chemistry. 2004;**69**:7092-7100. DOI: 10.1021/jo049375j

[16] Ryu DH, Kim KH, Sim JY, Corey EJ. Catalytic enantioselective Diels–Alder reactions of furans and 1,1,1-trifluoroethyl acrylate. Tetrahedron Letters. 2007;**48**:5735-5737. DOI: 10.1016/j.tetlet.2007.06.097

[17] Shibatomi K, Kobayashi F, Narayama A, Fujisawa I, Iwasa S. A Diels–Alder approach to the enantioselective construction of fluoromethylated stereogenic carbon centers. Chemical Communications. 2012;**48**:413-415. DOI: 10.1039/c1cc15889a

[18] Reddy KM, Bhimireddy E, Thirupathi B, Breitler S, Yu S, Corey EJ. Cationic chiral fluorinated oxazaborolidines. More potent, second-generation catalysts for highly enantioselective cycloaddition reactions. Journal of the American Chemical Society. 2016;**138**:2443-2453. DOI: 10.1021/jacs.6b00100

[19] Boobalan R, Chein R-J. Oxathiaborolium-catalyzed enantioselective [4 + 2] cycloaddition and its application in Lewis acid coordinated and chiral Lewis acid catalyzed [4 + 2] cycloaddition. Organic Letters. 2021;**23**:6760-6764. DOI: 10.1021/acs.orglett.1c02345

[20] Mukherjee S, Scopton AP, Corey EJ. Enantioselective pathway for the synthesis of laurenditerpenoly. Organic Letters. 2010;**12**(8):1836-1838. DOI: 10.1021/ol1004802

[21] Jung ME, Im G-YJ. Total synthesis of racemic laurenditerpenol, an HIF-1 inhibitor. Journal of Organic Chemistry. 2009;**74**:8739-8753. DOI: 10.1021/jo902029x

[22] Shingh RS, Adachi S, Tanaka F, Yamauchi T, Inui C, Harada T. Oxazaborolidinone-catalyzed enantioselective Diels-Alder reaction of acyclic α,β-unsaturated ketones. Journal of Organic Chemistry. 2008;**73**:212-218. DOI: 10.1021/jo702043g

[23] Davies HML, Ahmed G, Churchill MR. Asymmetric synthesis of highly functionalized 8-oxabicyclo [3.2.1]octene derivatives. Journal of the American Chemical Society. 1996;**118**: 10774-10782. DOI: 10.1021/ja962081y

[24] Olson JP, Davies HML. Asymmetric [4+3] cycloadditions between benzofuranyldiazoacetates and dienes: Formal synthesis of (+)-frondosin B. Organic Letters. 2008;**10**(4):573-576. DOI: 10.1021/ol702844g

[25] Villar L, Uria R, Martinez JI, Prieto L, Reyes E, Carrillo L, et al. Enantioselective oxidative (4+3) cycloadditions between allenamides and furans through bifunctional hydrogen-bonding/ion-pairing interactions. Angewandte Chemie International Edition. 2017;**56**:10535-10538. DOI: 10.1002/anie.201704804

[26] Banik SM, Levina A, Hyde AM, Jacobsen EN. Lewis acid enhancement by hydrogen-bond donors for asymmetric catalysis. Science. 2017;**358**: 761-764. DOI: 10.1126/science.aao5894

[27] Lehner V, Davies HML, Reiser O. Rh (II)-catalyzed cyclopropanation of furans and its application to the total synthesis of natural product derivatives. Organic Letters. 2017;**19**:4722-4725. DOI: 10.1021/acs.orglett.7b02009

[28] Hedley SJ, Ventura DL, Dominiak PM, Nygren CL, Davies HML. Investigation into factors influencing stereoselectivity in the reactions of heterocycles with donor-acceptor-substituted rhodium carbenoids. Journal of Organic Chemistry. 2006;**71**:5349-5356. DOI: 10.1021/jo060779g

[29] Olah GA, Krishnamurti R, Prakashi GKS. In: Trost BM, Fleming I, editors. Comprehensive Organic Synthesis. Vol. Vol. 3. Oxford: Pergamon Press; 1991. p. 293. DOI: 10.1016/B978-0-08-052349-1.00065-2

[30] Poulsen TB, Jørgensen KA. Catalytic asymmetric Friedel–Crafts alkylation reactions—Copper showed the way. Chemical Reviews. 2008;**108**(8): 2903-2915. DOI: 10.1021/cr078372e

[31] Gathergood N, Zhuang W, Jørgensen KA. Catalytic enantioselective Friedel-Crafts reactions of aromatic compounds with glyoxylate: A simple procedure for the synthesis of optically active aromatic mandelic acid esters. Journal of the American Chemical Society. 2000;**122**:12517-12522. DOI: 10.1021/ja002593j

[32] Zhung W, Gathergood N, Hazell RG, Jørsgensen KA. Catalytic, highly enantioselective Friedel-Crafts reactions of aromatic and heteroaromatic compounds to trifluoropyruvate. A simple approach for the formation of optically active aromatic and heteroaromatic hydroxy trifluoromethyl esters. Journal of Organic Chemistry. 2001;**66**:1009-1013. DOI: 10.1021/jo001176m

[33] Yamazaki S, Kashima S, Kuriyama T, Iwata Y, Morimoto T, Kakiuchi K. Enantioselective Friedel–Crafts reactions of ethenetricarboxylates and substituted pyrroles and furans and intramolecular reaction of benzene derivatives. Tetrahedron: Asymmetry. 2009;**20**: 1224-1234. DOI: 10.1016/j.tetasy.2009.05.016

[34] Kwiatkowski P, Wojaczy'nska E, Jurczak J. Asymmetric Friedel–Crafts reaction of furans with alkyl glyoxylates catalyzed by (salen)Co(II) complexes. Journal of Molecular Catalysis A: Chemical. 2006;**257**:124-131. DOI: 10.1016/j.molcata.2006.05.038

[35] Majer J, Kwiatkowski P, Jurczak J. Highly enantioselective synthesis of 2-furanyl-hydroxyacetates from furans via the Friedel-Crafts reaction. Organic Letters. 2008;**10**(14):2955-2958. DOI: 10.1021/ol800927w

[36] Oi S, Kashiwagi K, Inoue Y. A cationic palladium(II) complex-catalyzed Diels-Alder reaction. Tetrahedron Letters. 1998;**39**(34): 6253-6256. DOI: 10.1016/S0040-4039(98)01288-X

[37] Doherty S, Knight JG, Mehdi-Zodeh H. Asymmetric carbonyl-ene and Friedel–Crafts reactions catalysed by Lewis acid platinum group metal complexes of the enantiopure atropisomeric biaryl-like diphosphine (S)-Me2-CATPHOS: A comparison with BINAP. Tetrahedron: Asymmetry. 2012; **23**:209-216. DOI: 10.1016/j.tetasy.2012.01.022

[38] Paras NA. Enantioselective Organocatalytic Friedel-Crafts Alkylations of Heterocycles and Electron-Rich Benzenes. Caltech THESIS, California: Institute of Technology; 2004. DOI: 10.7907/EFG2-7V87

[39] Adachi S, Tanaka F, Watanabe K, Watada A, Harada T. Enantioselective Friedel–Crafts alkylation of furans and indoles with simple α,β-unsaturated ketones catalyzed by oxazaborolidinone. Synthesis. 2010;**15**:2652-2669. DOI: 10.1055/s-0029-1218821

[40] Kashikura W, Itoh J, Mori K, Akiyama T. Enantioselective Friedel–Crafts alkylation of indoles, pyrroles, and furans with trifluoropyruvate catalyzed by chiral phosphoric acid. Chemistry, an Asian Journal. 2010;**5**: 470-472. DOI: 10.1002/asia.200900481

[41] Yanagisawa A, Ishiba A, Nakashima H, Yamamoto H. Enantioselective addition of methallyl- and crotyltins to aldehydes catalyzed by BINAP·Ag(I) complex. Synlett. 1997;**1**: 88-90. DOI: 10.1055/s-0030-1258388

[42] Hanawa H, Hashimoto T, Maruoka K. Bis(((S)-binaphthoxy)(isopropoxy)titanium) oxide as a í-oxo-type chiral Lewis acid: Application to catalytic asymmetric allylation of aldehydes. Journal of the American Chemical Society. 2003;**125**:1708-1709. DOI: 10.1021/ja020338o

[43] Evans DA, Downey CW, Hubbs JL. Ni(II) Bis(oxazoline)-catalyzed enantioselective Syn aldol reactions of N-propionylthiazolidinethiones in the presence of silyl triflates. Journal of the American Chemical Society. 2003;**125**: 8706-8707. DOI: 10.1021/ja035509j

[44] Allais C, Nuhant P, Roush WR. (Diisopinocampheyl)borane-mediated reductive aldol reactions of acrylate esters: Enantioselective synthesis of *anti*-aldols. Organic Letters. 2013; **15**(15):3922-2935. DOI: 10.1021/ol401679g

[45] Nitabaru T, Nojiri A, Kobayashi M, Kumagai N, Shibasaki M. *anti*-Selective catalytic asymmetric nitroaldol reaction via a heterobimetallic heterogeneous catalyst. Journal of the American Chemical Society. 2009;**131**: 13860-13869. DOI: 10.1021/ja905885z

[46] Uraguchi D, Sakaki S, Ooi T. Chiral tetraaminophosphonium salt-mediated asymmetric direct henry reaction. Journal of the American Chemical Society. 2007;**129**:12392-12393. DOI: 10.1021/ja075152+

[47] Dangel BD, Polt R. Catalysis by amino acid-derived tetracoordinate complexes: Enantioselective addition of dialkylzincs to aliphatic and aromatic aldehydes. Organic Letters. 2000; **2**(19):3003-3006. DOI: 10.1021/ol0063151

[48] Hatano M, Miyamoto T, Ishihara K. 3,3′-Diphosphoryl-1,1′-bi-2-naphthol-Zn(II) complexes as conjugate acid-base catalysts for enantioselective dialkylzinc addition to aldehydes. The Journal of Organic Chemistry. 2006;**71**:6474-6484. DOI: 10.1021/jo060908t

[49] Fernández-Mateos E, Maciá B, Yus M. Catalytic enantioselective addition of organoaluminum reagents to aldehydes. Tetrahedron: Asymmetry. 2012;**12**:789-794. DOI: 10.1016/j.tetasy. 2012.05.007

[50] Yamamoto Y, Kurihara K, Miyaura N. Me-bipam for enantioselective ruthenium(II)-catalyzed arylation of aldehydes with arylboronic acids. Angewandte Chemie International Edition. 2009;**48**: 4414-4416. DOI: 10.1002/anie.2009 01395

[51] Fernández Ibáñez Ibanez MÁ, Macia B, Minnaard AJ, Feringa BL. Catalytic enantioselective reformatsky reaction with aldehydes. Angewandte Chemie International Edition. 2008;**47**: 1317-1319. DOI: 10.1002/anie.2007 04841

[52] Usanov DL, Yamamoto H. Asymmetric Nozaki–Hiyama propargylation of aldehydes: Enhancement of enantioselectivity by cobalt co-catalysis. Angewandte Chemie International Edition. 2010;**49**: 8169-8172. DOI: 10.1002/anie.2010 02751

[53] Usanov DL, Yamamoto H. Enantioselective alkynylation of aldehydes with 1-haloalkynes catalyzed by tethered bis(8-quinolinato) chromium complex. Journal of the American Chemical Society. 2011;**133**: 1286-1289. DOI: 10.1021/ja054871q

[54] Joannou MV, Moyer BS, Meek SJ. Enantio- and diastereoselective synthesis of 1,2-hydroxyboronates through Cu-catalyzed additions of alkylboronates to aldehydes. Journal of the American Chemical Society. 2015; **137**:6176-6179. DOI: 10.1021/jacs. 5b03477

Chapter 2

Density Functional Theory and Molecular Modeling of the Compound 2-[2-(4-Methylphenylamino)-4-phenylthiazol-5-yl]benzofuran

Yardily Amose, Fathima Shahana and Abbs Fen Reji

Abstract

The compound 2-[2-(4-methylphenylamino)-4-phenylthiazol-5-yl]benzofuran was prepared from 1-(4-methylphenyl)-3-(*N*-phenylbenzimidoyl)thiourea and 2-(2-bromoacetyl) benzofuran in the presence of triethylamine and characterized by FTIR, NMR, and mass spectra. Density functional theory (DFT) computations were adopted for the geometry optimization of this compound, to evaluate their Mulliken atomic charge distribution, HOMO-LUMO energy gap, and vibrational analysis. The titled compound induced G1 cell cycle arrest, which is regulated by CDK2 in cancer cells. Therefore, we used molecular modeling to study in-silico for the possible inhibitory effect as a mechanism of this compound as anticancer agents (PDB code: 2KW6, 6DL7, 6VJO, 6WMW, and 7LAE). The molecular docking study revealed that the compound was the most effective in inhibiting CDk2 cancer cells.

Keywords: benzofuran, DFT, vibrational analysis, molecular docking, anticancer agent

1. Introduction

Natural products have the potential to provide medicine with a source of novel structures. Nature is capable of producing complex molecules with numerous chiral centers that are planned to interact with biological systems. The marine environment is a rich source of biologically active natural products, many of which have not been originated in terrestrial sources [1, 2]. Marine natural products have fascinated the attention of biologists and chemists all over the world. As a consequence of the potential for new drug discovery, marine natural products have attracted scientists from different disciplines such as organic chemistry, bioorganic chemistry, pharmacology, biology, and ecology. From the studies 2,4-diaminothiazoloylbenzofuran and 2-aminothiazoloylbenzofuran analogs of dendrodoine have good docking characteristics, antimicrobial activities, we further planned to synthesize and evaluate the biological properties of 2-[2-(4-methylphenylamino)-4-phenylthiazol-5-yl]benzofuran (**Figure 1**) as further analogs of dendrodoine. These observations show that synthesis of

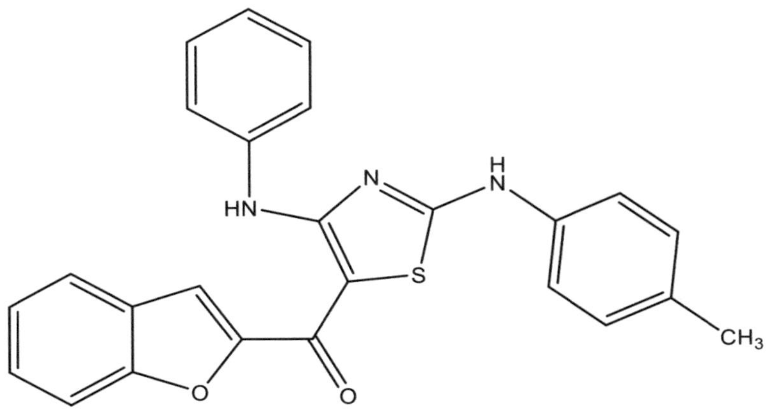

Figure 1.
Structure of 2-[2-(4-methylphenylamino)-4-phenylthiazol-5-yl]benzofuran.

2-[2-(4-methylphenylamino)-4-phenylthiazol-5-yl]benzofuran with a view to studying their biological activity, they exhibit a variety of bioactivity such as antibiotics, anticancer, anti-inflammatory, antitumor, antiviral, antibacterial, and antifungal activities. Hence, in this work the computational DFT calculation, particularly those based on hybrid functional method evolved to a powerful quantum chemical tool for the determination of the electronic structure of the molecule. Besides, molecular docking studies were carried out and the mechanisms of action of this compound on CDK2 cancer cell lines were studied.

2. Experimental

2.1 Material and methods

All chemicals were purchased from Sigma-Aldrich and were used without purification. It includes benzonitrile, aniline, anhydrous aluminum chloride, sodium hydroxide, triethylamine, *p*-tolyl isothiocyanate, and 2-bromoacetylbenzofuran. The organic solvents (spectroscopic grade) were used as received. The spectra had been documented on Bruker Avance400 FTNMR spectrometer (400 MHz for ^{1}H and ^{13}C NMR spectra), mass spectrometer on Agilent 6520(QTOF) positive mode ESI-MS, and Nicolet 400 FTIR spectrometer. The melting point was examined using digital melting point apparatus and uncorrected.

The density functional theory (DFT) was performed with Guassian-03 B3LYP/6-31G(d,p) basis set. Docking studies were carried out using the Hex 8.0 dock software with a grid dimension of 0.6. Discovery studio 3.5 visualizer was used to analyze the docking results.

2.2 General procedure for the synthesis of 2-[2-(4-methylphenylamino)-4-phenylthiazol-5-yl]benzofuran

To a solution of 1-aryl-3-(*N*-phenylbenzimidoyl)thiourea (1 mmol) in 5 ml *N,N*-dimethylformamide (1 mmol) was added. The mixture was stirred well and kept at room temperature for 5 hours. Triethylamine (2 mmol) was then added and the mixture was heated carefully at 55°C for 1 hour with occasional stirring afforded yellow precipitate. It was subsequently purified by crystallization from ethanol–water [3, 4].

2.3 Synthesis of 2-[2-(4-methylphenylamino)-4-phenylthiazol-5-yl]benzofuran

The orange yellow precipitate obtained was recrystallized using 2:1 ethanol–water solution. Yield 65.5%, m.p. 244–247, Analysis found: C, 73.63: H, 4.39: N, 7.02%: Calc. for $C_{25}H_{18}N_2O_2S$ (410.49): C, 73.15: H, 4.42: N, 6.82%: IR (KBr) cm-1: 3584, 3577, 3561, 3493, 3425, 3407, 3286, 3224, 3130, 3062, 3037, 3010, 2924, 2372, 1566, 1552, 1533, 1514, 1447, 1427, 1251, 1118, 1045, 1020, 746, 661. 1H NMR: (400 MHz, DMSO-d6) 2.37(s, 3H, CH3), 6.87 (d, 8.4 Hz, 2H, 2ArH), 7.18–7.39 (m, 7H H-5, H-6, 5ArH), 7.49–7.65(m, 4H, H-3, H-4, 2ArH), 7.87(d, 7.6 Hz, H-7), and 10.98(s, 1H, NH).

3. Results and discussion

3.1 Computational chemistry

3.1.1 Molecular geometry

The quantum chemical calculation is performed by DFT method with Becke's three parameters hybrid functional for the exchange part and the Lee-Yang-Parr (B3LYP) correlation function with 6-31G(d,p) basis set using Gaussian 09 program [5]. The optimized structure of the titled compound is depicted in (**Figure 2**). The optimized structure acquired structural parameters such as bond distance, angles, and dihedral angles are calculated [6–8].

3.1.2 Mulliken atomic charge distribution

The calculations of atomic charges explain the changes in dipole moment, molecular electronic structure as well as molecular polarizability. The partial atomic charges are a useful part of quantum mechanical calculation The calculated atomic charge values are taken from the B3LYP/6-31G(d,p) method. This calculation depicts the charges of all atoms in the titled compound. The Mulliken atomic charge of all hydrogen atoms is positive, all nitrogen and oxygen possess a negative charge and all sulfur carry a positive charge (**Figure 3**).

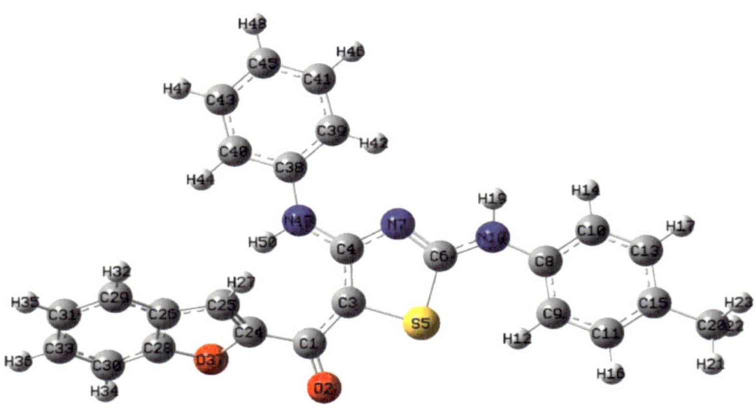

Figure 2.
Optimized structure of the compound 2-[2-(4-methylphenylamino)-4-phenylthiazol-5-yl] benzofuran.

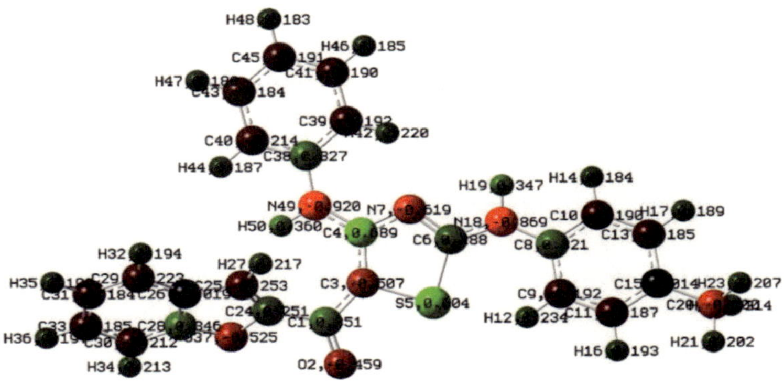

Figure 3.
Mulliken charge distribution of the compound 2-[2-(4-methylphenylamino)-4-phenylthiazol-5-yl]benzofuran.

3.1.3 Analysis of frontier molecular orbitals

HOMO-LUMO energy gap explains the chemical reactivity of the molecule. If the energy gap is less, it is more reactive and if it is high, the compound is thermally stable [9]. The thermal stability of the compound is related to the hardness of the molecule. It is found that the charge distribution of the HOMO level of the titled compound is mostly localized on the thiazole and phenyl rings and the charge

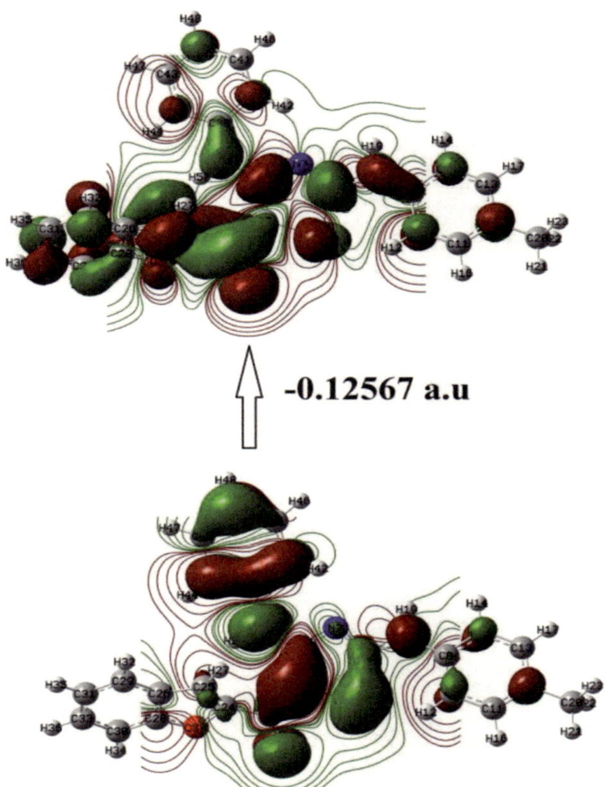

Figure 4.
HOMO-LUMO energy diagram of the compound 2-[2-(4-methylphenylamino)-4-phenylthiazol-5-yl] benzofuran.

distribution of the LUMO level is delocalized throughout the molecule. The energy gap is found to be less than −0.1256 a.u (**Figure 4**).

3.1.4 Vibrational analysis

The spectroscopic signature of the titled compound was performed by FT-IR spectra. The theoretical vibrational frequency of the compound was calculated using the B3LYP/6-31G method. The titled compound consists of 50 atom that produces 144 normal modes of vibrations.

The bands at 3497 cm^{-1} are due to the N-H stretching vibration of the secondary amine. The bands at 3125 cm^{-1}, 3096 cm^{-1} are due to the C-H stretching vibration. The bands at 1637 cm^{-1} are due to the C=O stretching vibration. The C-N stretching modes were observed in 1554 cm^{-1} (**Figure 5**) [9, 10].

3.1.5 Molecular docking

HEX is an interactive molecular graphics program for calculating and displaying feasible docking modes among the protein and the DNA molecules. To find out the antibacterial activity and binding energy of the titled compound, the molecule should bring to minimized energy level using 6-31 g(d,p) software system, and also the compound should obey the Lipinski rule of five shown in (**Table 1**). The molecule is docked into the active site of the CDK2 in cancer cells (PDB code: 2KW6, 6DL7, 6VJO, 6WMW, and 7LAE). Docking results were analyzed based on binding energy and hydrogen bonding [11, 12]. The correct interaction conformation between ligand and protein receptor is explained by the π–σ, π–cation, π–π interaction and Van der Wall interaction (**Figure 6** and **Table 2**). Based on the results, it is clear that the compound binds favorably with the protein receptor.

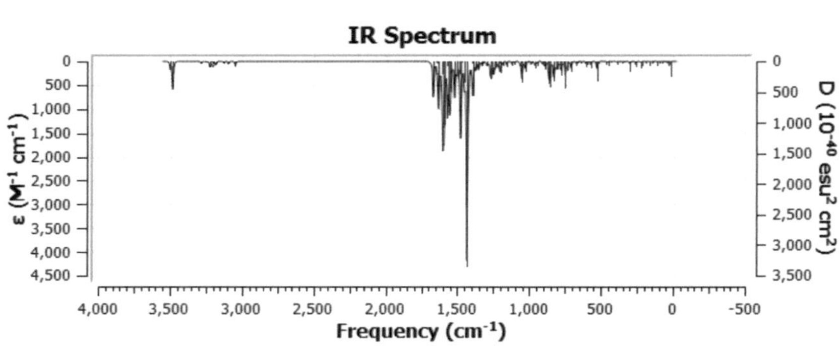

Figure 5.
Calculated IR spectrum of 2-[2-(4-methylphenylamino)-4-phenylthiazol-5-yl]benzofuran.

Compound	Molecular weight (<500 Da)	HB donar (<5)	HB acceptor (<10)	Log P < 5	Molecular refractivity (40–130)
2-[2-(4-methylphenylamino)-4-phenylthiazol-5-yl] benzofuran	425	2	5	6.91	125.94

Table 1.
LIPINSKI RULE OF 2-[2-(4-methylphenylamino)-4-phenylthiazol-5-yl]benzofuran.

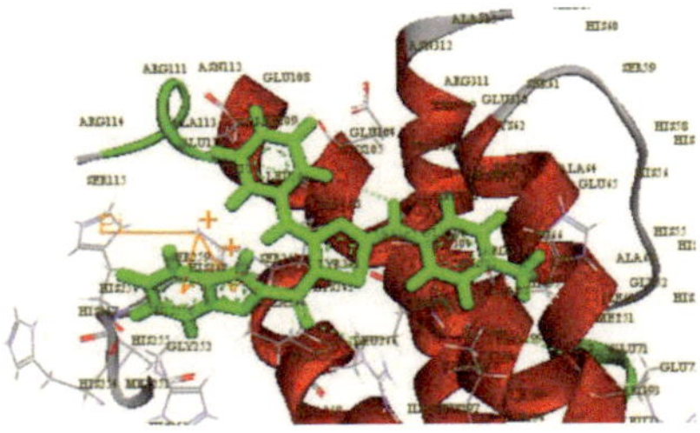

(i)

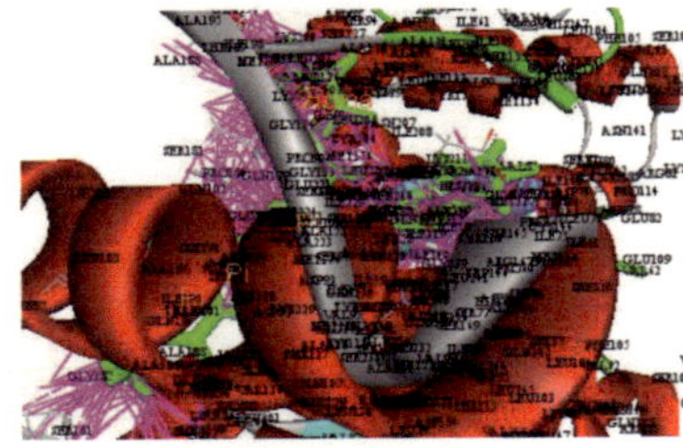

(ii)

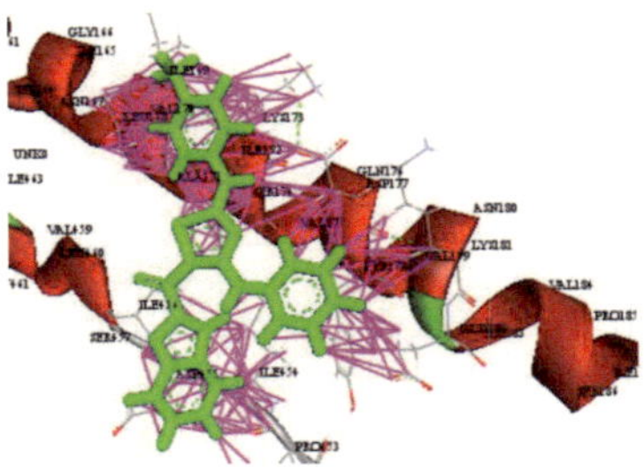

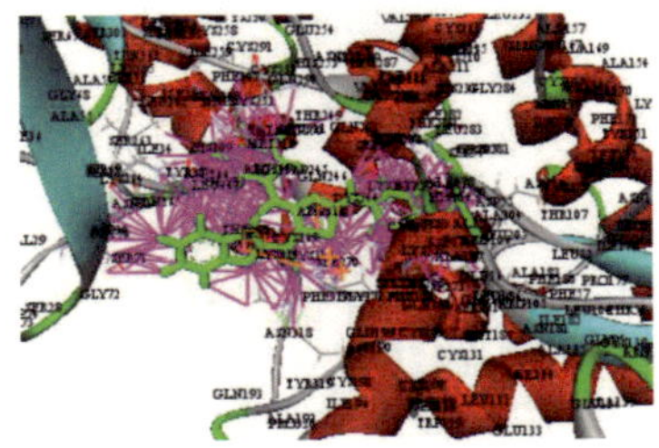

(iii)

(iv)

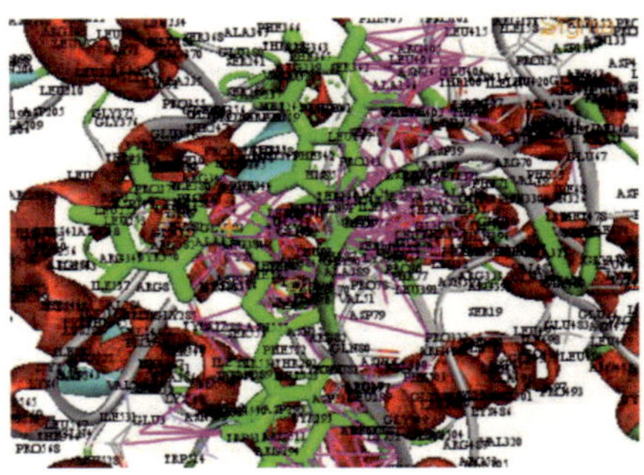

(v)

Figure 6.
(i)–(v) 3D docking structure of the titled compound with the protein receptors 2KW6, 6DL7, 6VJO, 6WMW, and 7LAE.

Cancer cell (PDB code)	Binding energy (kJ/mol)	Active sites of interactions				
		π–σ interactions	π–cation interactions	π–π interaction	Electrostatic	Van der Waals
2KW6	−294.25	—	ARG A:103	—	ARG A:103, ARG A:99, GLY A:100	HIS B:254, HIS A:97, GLY B:252
6DL7	−356.35	ILE B:59	—	TYR B:76	ASP B:74, SER B:77, TYR B:76	ILE B:59
6VJO	−287.89	—	—	—	TYR B:178, VAL B:170, ILE A:454	ASP B: 177, ASP A:455
6WMW	−270.58	—	ARG B:313, LYS B:314, LYS B:202	—	HIS B:206,ARG B:313, LYS B:314, LYS B:202	TYR B:175, LEU B:186, GLY L:33, ALA B: 187

Cancer cell (PDB code)	Binding energy (kJ/mol)	Active sites of interactions				
		π–σ interactions	π–cation interactions	π–π interaction	Electrostatic	Van der Waals
7LAE	329.14	—	—	CYS D1	CYS D1, GLU A:38, ASP A:39	VAL A:51, THR A:48, TYR A:45

Table 2.
Docking score and interaction of the compounds with cancer cell line.

4. Conclusion

Benzofuran derivatives have a broad spectrum of biological activities such as antimicrobial, antifungal, anti-inflammatory, anticancer, and analgesic and it is understood that many natural products with benzofuran moiety exhibit interesting biological and pharmacological activities. We have established the modest synthetic techniques of benzofuran analogs of dendrodoine *viz.* 2-[2-(4-methylphenylamino)-4-phenylthiazol-5-yl]benzofuran and characterized by IR, ^{1}H NMR, ^{13}CNMR, and mass spectra. Theoretical information on the optimized geometry, atomic charges, and frontier molecular orbitals in the ground state were obtained using density functional theory (DFT) using standard B3LYP/6-31G basis sets with Gaussian '09 software. Mulliken population analysis was performed on the atomic charges distribution and the HOMO-LUMO energies were calculated and found that the compound is more reactive which is clearly shown in the docking study. The compound was docked with five CDK2 cancer cells. Among them, the cancer cell with PDB code 6DL7 binds more favorable with the titled compound and shows relative binding energy of −356.35 kcal/mol.

Conflict of interest

All authors declare no conflict of interest.

Author details

Yardily Amose[1*], Fathima Shahana[1] and Abbs Fen Reji[2]

1 Department of Chemistry and Research Centre, Scott Christian College, Nagercoil, Tamil Nadu, India

2 Department of Chemistry, Nesamony Memorial Christian College, Kanyakumari, Tamil Nadu, India

*Address all correspondence to: ayardily@gmail.com

References

[1] Brad KC. Biomedical Potential of Marine Natural Product. Bioscience. 1996;**46**:271-286

[2] Thomas KK, Reshmy R, Ushadevi KS. Synthesis of a novel bioactive 2-substituted amino-5-indol-3-oyl-4-phenylthiazoles. Journal of the Indian Chemical Society. 2007;**84**:1016-1019

[3] Alwin T, Abbs Fen Reji TF. Synthesis, antioxidant and antibacterial studies on 2-(2-arylamino-4-phenylthiazol-5-yl) benzofuran derivatives. International Research Journal of Pure and Applied Chemistry. 2017;**15**(1):1-8

[4] Silverstein RM, Bassler GC, Morrill TC. Spectrometric Identification of Organic Compounds. 4th ed. New York, USA: John Wiley & Sons, Inc.; 1981

[5] Ganesan MS, Kanmani Raja K, Murugesan S, Kumar BK, Rajagopal G, Thirunavukkarasu S. Synthesis, biological evaluation, molecular docking, molecular dynamics and DFT studies of quinoline-fluoroproline amide hybrids. Journal of Molecular Structure. 2020;**1217**:128360

[6] Becke AD. Density-functional thermo chemistry. III. The role of exact exchange. The Journal of Chemical Physics. 1993;**98**:5648

[7] Khalid M, Ullah MA, Adeel M, Khan MU, Tahir MN, Braga AAC. Synthesis, crystal structure analysis, spectral IR, UV-Vis, NMR assessment, electronic and non-linear optical properties of potent quinolone based derivatives: Interplay of experimental and DFT study. Journal of Saudi Chemical Society. 2019;**23**(5):546-560

[8] Boukabcha N, Benhalima N, Rahmani R, Chouaih A, Hamzaoui F. Theoretical investigation of electrostatic potential and non-liner optical properties of m-nitroacetanilide. Rasayan Journal of Chemistry. 2015;**8**(4):509-516

[9] Kecel-Gunduz S, Bicak B, Celik S, Akyuz S, Ozel AE. Structural and spectroscopic investigation on antioxidant dipeptide, l-methionyll-serine: A combined experimental and DFT study. Journal of Molecular Structure. 2017;**1137**:756-770

[10] Celik S, Akyuz S, Ozel AE. Vibrational spectroscopic and structural investigations of bioactive molecule glycyl-tyrosine (Gly-Tyr). Vibrational Spectroscopy. 2017;**92**:287-297

[11] Shahana MF, Yardily A. Synthesis, spectral characterization, DFT, and docking studies of (4-amino-2-(phenylamino) thiazol-5-yl)(thiophene-2-yl)methanone and (4-amino-2-(4-chlorophenyl)(amino) thiazol-5-yl)(thiophene-2-yl) methanone. Journal of Structural Chemistry. 2020;**61**:1367-1379

[12] Shahana MF, Yardily A. Synthesis, quantification, DFT calculation and molecular docking of (4-amino-2-(4-methoxyphenyl)aminothiazol-5yl)(thiophene-2-yl)methanone. Indian Journal of Biochemistry and Biophysics. 2020;**57**:606-612

Chapter 3

Synthesis and Characterization of New Racemic α-Heterocyclic α,α-Diaminoester and α,α-Diamino Acid Carboxylic: 2-Benzamido-2-[(Tetrahydro-Furan-2-Ylmethyl) Amino]Acetate and 2-Benzamido-2-[(Tetrahydro-Furan-2-Ylmethyl) Amino] Acetic Acid

El Houssine Mabrouk

Abstract

We reported here the synthesis of new α,α-diaminoester and α,α-diamino acid derivatives, as 2-benzamido-2-[(tetrahydro-furan-2-ylmethyl)amino] acetic acid through alkaline hydrolysis reaction of corresponding N-benzoylated methyl α,α-diamino ester. The α,α-diaminoester derivative was synthesized by nucleophilic substitution of methyl α-azido glycinate N-benzoylated with 2-tetrahydrofuran-2-ylmethan-amine. The structure of these products were established on the basis of NMR spectroscopy (^{1}H, ^{13}C), and MS data.

Keywords: Amino acid, Amine, Heterocyclic molecules, Nucleophilic substitution, Methyl α-azido glycinate, α,α-Diamino ester, α,α-diamino acid, alkaline hydrolysis reaction

1. Introduction

α-amino acids are a considerable interest due to their diversity in several fields of research (asymmetric organic synthesis and medicine) and applications (food industry and drugs). Given the very broad activity they present, research teams are interested in evaluating and developing their role for very effective use [1–5]. This has led to the development of numerous synthetic methods for a variety of compounds [6]. Heterocyclic chemistry is the basis for the discovery of the importance of the multiple medical properties of these compounds [7–12].

Because of their multiple functionalities, heterocyclic amino acids play a considerable role in the biologic processes [13, 14]. Therefore, a large number of between them isolated of plants have a very varied biologic activity [15].

Heterocyclic compounds have a wide spectrum activities, including antimicrobial [16, 17] and antibacterial properties [18, 19], anticancer agents [20], antiviral [21], antitumor activity [22], and in agricultural science as potent fungicides, herbicides and insecticides [23]. Heterocyclic amino acids and their derivatives represent a well-known group of organic compounds also presenting biological activity [24–26]. Considering the interest in these heterocyclic amino acids, several structurally related nonproteinogenic amino acids and their derivatives have been the subject of various investigations [27–29]. We present herein a convenient and easy procedure for the preparation of new racemic carboxylic α,α-diamino acid derivative with the aim to have access to new active biomolecule with a good yield.

For this reason, we considered it interesting to synthesize new compounds containing 2-tetrahydrofuran-2-ylmethanamine [30, 31] fused with an amino acid, in order to study their biological activities. The present study describes the synthesis and characterization of new α,α-diamino ester and α,α-diamino acid derivatives.

2. Results and discussions

2.1 Synthesis of new racemic α-carboxylic α,α-diamino ester

2.1.1 Preparation of azide dipole

The first step of the synthesis strategy that we have developed consists in preparing the methyl ester of glycine N-protected by benzoyl chloride. Thus, the esterification of the starting amino acid is carried out by the action of thionyl chloride in anhydrous methanol on glycine and leads to the corresponding hydrochloride **1** with good yield (Yield = 92%). The hydrochloride is neutralized by adding triethylamine or bubbling gaseous ammonia (**Figure 1**).

The methyl glycinate **2** thus prepared is protected with benzoyl chloride. The protection reaction is carried out in dichloromethane in the presence of triethylamine or pyridine. After chromatography on a silica gel column, the N-protected aminoester **3** is obtained in good yields. Different protecting groups may protect the methyl glycinate **2**: trifluoroacetic anhydride, trichloroethoxycarbonyl chlorides and acetyl chloride.

The bromination reaction of N-protected amino ester is carried out by bromine in the presence of α,α'-azo-bis-isobutyronitrile (AIBN) in a catalytic amount or by N-bromosuccinimide, in reflux of carbon tetrachloride and under the irradiating action of a 300 W lamp. In general, methyl N-benzoyl α-bromoglycinate **4** is obtained in excellent yields. This product is used in most cases without purification in the next step (**Figure 1**).

The substitution of bromine by the azide group is effected by the action of sodium azide, in acetone, at room temperature for a period ranging from 4 to 5 hours. Purification of the reaction crude by chromatography on a silica gel column (eluent: 50/50 ether/hexane) allows substitution product **5** to be obtained with excellent yield. The title compound is stable and can be stored for an unlimited time without any signs of decomposition. The methyl α-bromo glycinate **4** also can be used and gives satisfactory results; the azide **5** is used especially for its stability.

2.1.2 N-alkylation of the amine by N-protected methyl α-azidoglycinate 5

We continued our investigations on the use of organic azides [30, 32–37] in heterocyclic synthesis; we reported in this paper another part of our investigations concerning the preparation of new carboxylic α,α-diaminoester carrying

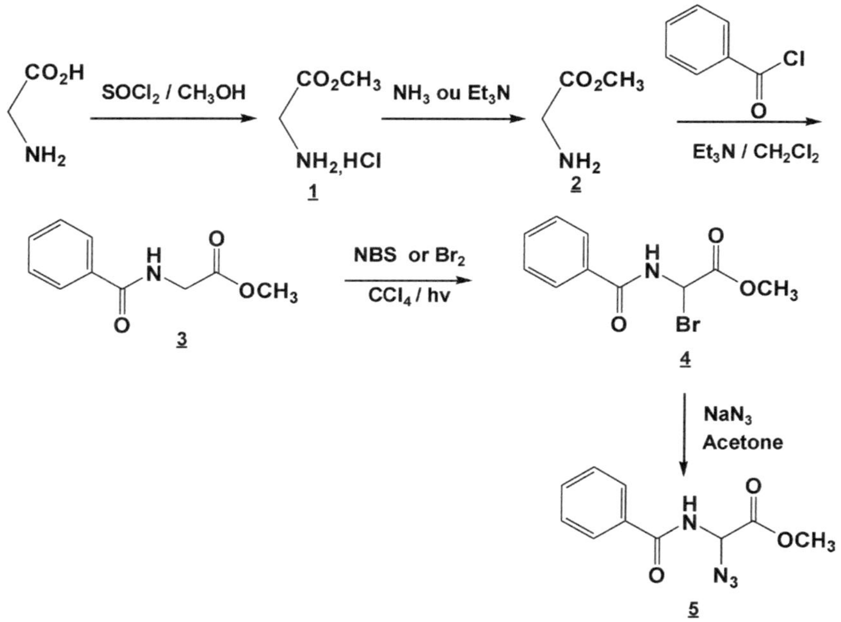

Figure 1.
Preparation of N-protected methyl α-azidoglycinate 5.

heterocyclic in position α. Our strategy is based on the nucleophilic substitution of methyl α-azido glycinate N-benzoylated **5** with 2-tetrahydrofuran-2-ylmethanamine (**Figure 2**).

In order to make a comparative study and to find the best experimental mode for this synthesis strategy, we carried out the synthesis reaction of the α,α-diaminoester in the absence and in the presence of a base (triethylamine (Et_3N) or diisopropylethylamine (DIEPA)) in a solvent (dichloromethane (DCM) or acetone). The results in **Table 1** clearly show that the good yield is obtained using acetone in the presence of DIEPA.

2.2 Synthesis of new racemic α-carboxylic α,α-diamino acid

In continuation of our research, we will present in this work, our results concerning the synthesis of new α,α-diamino acid derivative, as 2-benzamido-2-[(tetrahydrofuran-2-ylmethyl)amino] acetic acid through alkaline hydrolysis reaction of corresponding *N*-benzoylated methyl α,α-diamino ester [31]. After the obtaining of the *N*-protected methyl α,α-diamino ester **6**, we proceeded to the cleavage of the protecting groups to obtain the corresponding α,α-diamino acid **7**. The hydrolysis reaction of the α,α-diamino ester methyl 2-benzamido-2-[(tetrahydro-furan-2-ylmethyl)

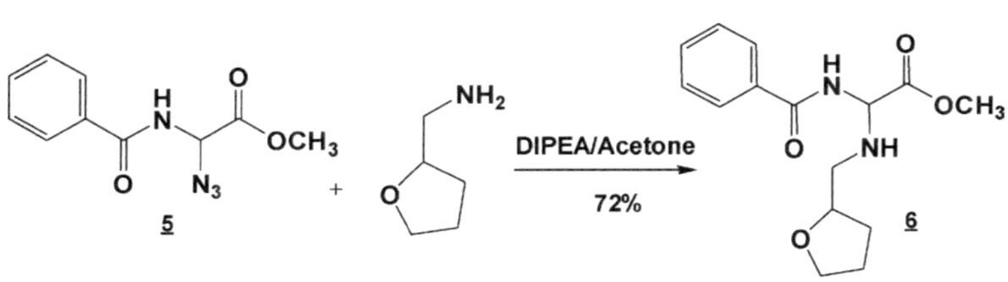

Figure 2.
2-Benzamido-2-[(tetrahydro-furan-2-ylmethyl) amino] methyl acetate ***6***.

Product	- DCM Yield (%)	Et_3N DCM Yield (%)	Et_3N Acetone Yield (%)	DIEPA DCM Yield (%)	DIEPA Acetone Yield (%)
6	Traces	17.5	30.5	39	72

Table 1.
Different operating conditions for the synthesis reaction of 2-benzamido-2 - [(tetrahydro-furan-2-ylmethyl) amino] methyl acetate 6.

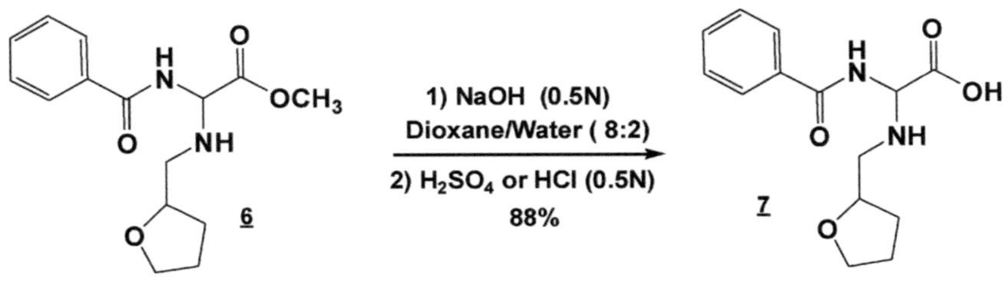

Figure 3.
2-Benzamido-2-[(tetrahydro-furan-2-ylmethyl)amino] acetic acid 7.

amino]acetate **6** in a basic medium is carried out for approximately 30 minutes and leads, after acidification of the reaction medium with sulfuric acid or hydrochloric acid, to the corresponding α,α-diamino acid 2-benzamido-2-[(tetrahydro-furan-2-ylmethyl) amino] acetic acid 7 in good yield (**Figure 3**).

3. Conclusion

The first step in our synthesis strategy is to prepare the azide dipole. In the second step, our objective is the preparation of carboxylic α,α-diaminoester and diamino acid carrying a heterocycle in position α. This method provides an easy procedure for the preparation of new carboxylic α,α-diamino acids derivatives in very satisfactory yields starting from the appropriate azide derivative **5**. The nucleophilic substitution of methyl α-azido glycinate N-benzoylated **5** by 2-tetrahydrofuran-2-ylmethanamine occurred under very mild conditions and led to the 2-benzamido-2-[(tetrahydro-furan-2-ylmethyl)amino]acetate in good yield. 2-benzamido-2-[(tetrahydro-furan-2-ylmethyl)amino] acetic acid was synthesized through alkaline hydrolysis reaction of corresponding N-benzoylated methyl α,α-diaminoester.

4. Experimental

4.1 *N*-alkylation reaction procedure

To facilitate the nucleophilic attack of 2-tetrahydrofuran-2-ylmethanamine ($5.72. 10^{-3}$ moles) on methyl α-azido glycinate ($5.2. 10^{-3}$ moles), one adds at the start ($6.24. 10^{-3}$ moles) of diisopropylethylamine on the amine in 20 ml of dry acetone. Deprotonation of the amine is carried out with stirring for one hour before adding the azide. At the end of the reaction which takes place for 48 hours, the evaporation of the solvent takes place under reduced pressure, decantation is ensured by dichloromethane or ethyl acetate using (Na_2SO_4) as desiccant, the product N-alkylated is purified by recrystallization or chromatography on a silica gel column (hexane/ether).

4.2 *N*-benzoylated methyl α,α-diaminoester: Methyl 2-benzamido-2-[(tetrahydro-furan-2-ylmethyl)amino]acetate 6

Yield = 72.0%; m. p. °C (hexane/ether (1/1)): 130–132; R_f (ether) = 0.63; M.S. (electrospray) m/z = 292.3 [M]; 293.3 [M + 1]; $C_{15}H_{20}N_2O_4$. ^{13}C NMR (δppm, $CDCl_3$): (2CO): 171.98, 169.07; C_6H_5 (aromatic): 135.77, 131.33, 129.54, 128.34; (OCH): 77.23; (-CH-): 71.32; (OCH_3): 54.49; 4· (CH_2): 66.12, 50.25, 27.07, 24.13. ^{1}H NMR (δppm, $CDCl_3$): 8.02–7.40 (5H, NH_{amid} + Ar, 3 m); 5.60 (1H, H_α, br s); 4.45–4.10 (3H, NH + H_{THF}, 2 m); 3.75 (3H, OCH_3, s); 2.90 (2H, NCH_2, m); 1.75–1.20 (5H, H_{THF}, 2 m).

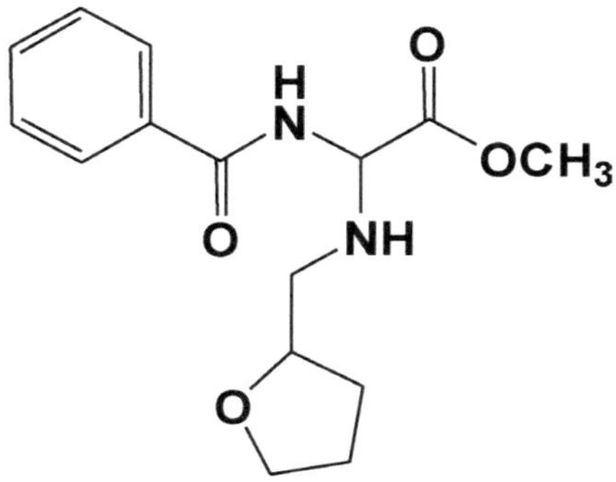

4.3 Deprotection of acid function: Synthesis of *N*-benzoylated α,α-diamino acid derivative 7

To a solution of the *N-benzoylated* α,α-diamino ester *derivative* (1 mmole) in 10 mL of dioxane/water mixture (8/2), one adds 1.5 mmole of NaOH (0.5 N) with stirring and at 0°C. The stirring is maintained at room temperature until disappearance of the starting material. The reaction is always followed by TLC. The solvent is then evaporated and the pH of the aqueous phase is adjusted to 6 using a solution of sulfuric acid or hydrochloric acid (0.5 N). One extracts with ethyl acetate and the organic layers recovered, are dried and concentrated under vacuum. The product is recrystallized from ether/hexane.

4.4 2-Benzamido-2-[(tetrahydro-furan-2-ylmethyl)amino]acetic acid 7

Yield: 88%; Rf: 0.7 (ether); ^{1}HNMR ($CDCl_3$): δppm: 1.25–1.80 (2 m, 5H, $H_{T.H.F}$); 2.85 (m, 2H, NCH_2); 4.20–4.50 (2 m, 3H, $H_{T.H.F}$ + NH); 5.5 (br s, 1H, H_α); 7.45–8.00 (3 m, 5H, Ar + N-H_{amid}). ^{13}C NMR ($CDCl_3$): δppm: 4 (CH_2) 25.45, 27.18, 52.05, 66.42; 72.12 (-CH-); 77.26 (OCH); 127.84, 129.34, 131.72, 135.68 (C_6H_5 aromatic carbons); 169.12, 171.92 (2CO). MS (electrospray) m/z = 279.2 [M + 1]; 278.2 [M]; $C_{14}H_{18}N_2O_4$.

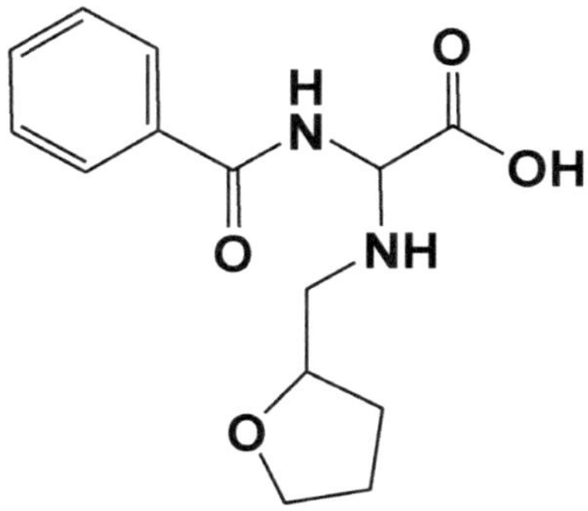

Author details

El Houssine Mabrouk
Laboratory of Materials Engineering for the Environment and Natural Resources, Faculty of Sciences and Technologies, Moulay Ismail University, Errachidia, Morocco

*Address all correspondence to: mabrouk.elhoussine@gmail.com; e.mabrouk@umi.ac.ma

References

[1] Beers SA, Schwender CF, Loughney D, Malloy E, Demarest K. Phosphatase inhibitors—III. Benzylaminophosphonic acids as potent inhibitors of human prostatic acid phosphatase. Bioorg. Med. Chem. 1996; 4: 1693-1701. Doi.org/10.1016/0968-0896(96)00186-1

[2] Mikolajczyk MJ. Organomet. Chem. 2005; 690: 2488.

[3] Joly GD, Jacobsen EN. Thiourea-catalyzed enantioselective hydrophosphonylation of imines: practical access to enantiomerically enriched α-amino phosphonic acids. J. Am. Chem. Soc. 2004; 126: 4102-4103. Doi.org/10.1021/ja0494398

[4] Leite AL, Lima RS, Moreira DRM, Cardoso MV, Brito ACG, Santos LMF, Hernandes MZ, Kiperstok AC, Limac RS, Soaresc MBP. Synthesis, docking, and in vitro activity of thiosemicarbazones, aminoacyl-thiosemicarbazides and acyl-thiazolidones against Trypanosoma cruzi. Bioorg. Med. Chem. 2006; 14: 3749-3757. Doi.org/10.1016/j.bmc.2006.01.034

[5] Moreira DRD, Leite ACL, Ferreira PMP, Da Costa PM, Costa Lotufo BL., De Moraes MO, Brondani DJ, O′ Pessoa CD. Synthesis and antitumour evaluation of peptidyl-like derivatives containing the 1, 3-benzodioxole system. Eur. J. Med. Chem. 2007; 42: 351-357. Doi.org/10.1016/j.ejmech.2006.10.007

[6] Haemers A, Mishra L, Vanassche I, Bollaert W. Asymmetric synthesis of aminoacids by enantio and diastereodifferentiating reactions. Die Pharm. 1989; 44: 97-144. http://pascal-francis.inist.fr/vibad/index.php?action=getRecordDetail&idt=6571647.

[7] Boschelli DH, Connor DT, Barnemeier DA. 1, 3, 4-Oxadiazole, 1, 3, 4-thiadiazole, and 1, 2, 4-triazole analogs of the fenamates: in vitro inhibition of cyclooxygenase and 5-lipoxygenase activities. J. Med. Chem. 1993; 36: 1802-1810. Doi.org/10.1021/jm00065a002

[8] Unangst PC, Shrum.P, Connor DT, Dyer DR, Schrier DJ. Novel 1, 2, 4-oxadiazoles and 1, 2, 4-thiadiazoles as dual 5-lipoxygenase and cyclooxygenase inhibitors. J. Med. Chem. 1992; 35: 3691-3698. Doi.org/10.1021/jm00098a015

[9] Hung J, Chen RY. Synthesis and bioactivities of 1,3,2-benzodiaza phosphorin-2-carboxamide 2-oxides containing α-aminophosphonate groups. Heteroat. Chem. 2001; 12: 97-101. Doi.org/10.1002/hc.6

[10] Srivastava N, Bahadur S, Verma HN, Khan MM. Synthesis of new 5, 3/5 and 2-substituted (1, 3, 4)-oxadiazoles and their related products as potential antifungal and antiviral agents. Curr. Sci. 1984; 53: 235-239. https://www.jstor.org/stable/24215988

[11] Saad H. Synthesis of some pyridyloxymethyloxadiazoles, thiadiazoles and triazoles of expected pharmalogical activity. Indian J. Chem. 1996; 35B: 980-984.

[12] Hui XP, Zang CH, Wang Q, Zhang Q. Synthesis and antibacterial activities of 1,3,4-oxadiazole derivatives containing 5-methylisoxazole moiety. Indian J. Chem. 2002; 41B: 2176-2179. http://hdl.handle.net/123456789/22074

[13] Costantino G, Marinozzi M, Camaioni E, Natalini B, Sarichelou I, Micheli F, Cavanni P, Faedo S, Noe C, Morini F, Pelliciari R. Stereoselective synthesis and preliminary evaluation of (+)- and (−)-3-methyl-5-carboxy-thien-2-yl-glycine (3-MATIDA): identification of (+)-3-MATIDA as a novel mGluR1

competitive antagonist. Ilfarmaco. 2004; 59: 93-99. Doi.org/10.1016/j.farmac.2003.11.008

[14] Jorgensen CG, Osborne HB, Nielsen B, Kehler J, Clausen RP, Krogsgaard-Larsena P, Madsena U. Novel 5-substituted 1-pyrazolol analogues of ibotenic acid: synthesis and pharmacology at glutamate receptors. Bioorg. Med. Chem. 2007; 15: 3524-3538. Doi.org/10.1016/j.bmc.2007.02.047

[15] Schenk SU, Werner D. β-(3-isoxazolin-5-on-2-yl)-alanine from Pisum: allelopathic properties and antimycotic bioassay. Photochemistry. 1991; 30: 467-470. Doi.org/10.1016/0031-9422(91)83706-Q

[16] Xu W, Song B, Bhadury P, Song Y, Hu D. Synthesis and Crystal Structure of Novel Sulfone Derivatives Containing 1,2,4-Triazole Moieties. Molecules, 2010; 15: 766-779. Doi.org/10.3390/molecules15020766

[17] Karabasanagouda T, Adhikari AV, Shetty NS. Synthesis and antimicrobial activities of some novel 1,2,4-triazolo [3,4-b]-1,3,4-thiadiazoles and 1,2,4-triazolo[3,4-b]-1,3,4-thiadiazines carrying thioalkyl and sulphonyl phenoxy moieties. Eur. J. Med. Chem. 2007; 42: 521-529. Doi.org/10.1016/j.ejmech.2006.10.010

[18] Tehranchian S, Akbarzadeh T, Fazeli MR, Jamalifar H, Shafiee A. Synthesis and antibacterial activity of 1-[1,2,4-triazol-3-yl] and 1-[1,3,4-thiadiazol-2-yl]-3-methylthio-6,7-dihydro benzo[c]thiophen-4(5H) ones. Biol. Med. Chem. Lett. 2005; 15: 1023-1025. Doi.org/10.1016/j.bmcl.2004.12.039

[19] Foroumadi A, Soltani F, Moshafi MH, AshrafAskari R. Synthesis and in vitro antibacterial activity of some N-(5-aryl-1,3,4-thiadiazole-2-yl) piperazinyl quinolone derivatives. Farmaco. 2003; 58: 1023-1028. Doi.org/10.1016/S0014-827X(03)00191-5

[20] Holla BS, Poorjary KN, Rao BS, Shivananda MK. New bis-aminomercaptotriazoles and bistriazolothiadiazoles as possible anticancer agents. Eur. J. Med. Chem. 2002; 37: 511-517. Doi.org/10.1016/S0223-5234(02)01358-2

[21] Yaseen AAS, Mohammad NAD, Najim AAM. Synthesis, antitumor and antiviral properties of some 1,2,4-triazole derivatives. Farmaco. 2004; 59: 775-783. Doi.org/10.1016/j.farmac.2004.05.006

[22] Demirbas N, Ugurluoglu R, Demirbas A. Synthesis of 3-alkyl(Aryl)-4-alkylidenamino-4,5-dihydro-1H-1,2,4-triazol-5-ones and 3-alkyl-4-alkylamino-4,5-dihydro-1H-1,2,4-triazol-5-ones as antitumor agents. Bioorg. Med. Chem. 2002; 10: 3717-3723. Doi.org/10.1016/S0968-0896(02)00420-0

[23] Song BA, Hu DY, Zheng S, Huang RM, Yang S, Huang J. Synthesis and bioactivity of phosphorodithioate compounds containing 1,2,4-triazole. Chin. J. Org. Chem. 2001; 21: 524-529.

[24] Ebert B, Lenz S, Brehem L, Bregnedal P, Hansen J, Frederiksen K, Bogeso K, KrosgardLarsen P. 1,2,3-Triazolyl Amino Acids as AMPA Receptor Ligands. J. Med. Chem. 1994; 37: 878-884.

[25] Ikegami F, Murakoshi I. Enzymic synthesis of non-protein b-substituted alanines and some higher homologues in plants. Phytochemistry. 1994; 35: 1089-1104. Doi.org/10.1016/S0031-9422(00)94805-2

[26] Dunnill P, Fowden L. The amino acids of seeds of the Cucurbitaceae. Phytochemistry. 1965; 4: 933-944. Doi.org/10.1016/S0031-9422(00)86271-8

[27] Ferreria PMT, Maia HLS, Monterio LS. High yielding synthesis of heterocyclic β-substituted alanine derivatives. Tetrahedron Lett. 1999; 40: 4099-4102. Doi.org/10.1016/S0040-4039(99)00691-7

[28] Hyslop JF, Lovelock SL, Watson AJB, Sutton PW, Roiban GD. N-Alkyl-α-amino acids in Nature and their biocatalytic preparation. Journal of Biotechnology. 2019; 293: 56-65. Doi.org/10.1016/j.jbiotec.2019.01.006

[29] Motta P, Pollegioni L, Molla G. Properties of lamino acid deaminase: En route to optimize bioconversion reactions. Biochimie. 2019; 158: 199-207. Doi.org/10.1016/j.biochi.2019.01.010

[30] Mabrouk EH, Elachqar A, El Hallaoui A, El Hajji S, Martinez J, Rolland V. Synthesis of new racemic α-heterocyclic α,α-diaminoesters and α-aminoester carboxylic. Arabian Journal of Chemistry. 2013; 6: 93-96. Doi:10.1016/j.arabjc.2010.09.023.

[31] Mabrouk EH. Synthesis of 2-benzamido-2-(3-amino-1,2,4-triazol-1-yl) acetic acid and 2-benzamido-2-[(tetrahydro-furan-2-ylmethyl) amino] acetic acid. IJCPS. 2019; 7(5): 65-68.

[32] Achamlale S, Mabrouk H, Elachqar A, El Hallaoui A, et al. Synthesis and thermal isomerization of carboxylic and phosphonic α-aminoesters substituted with a triazole ring, Phosphorus, Sulfur Silicon Relat. Elem. 2007; 182: 357-367. Doi.org/10.1080/10426500600919959

[33] Mabrouk EH, Elachqar A, El Hallaoui A. Synthesis of new racemic α,α-diamino carboxylic ester derivatives. Molecules. 2010; 13: 9354-9363. Doi.org/10.3390/molecules15129354

[34] Mabrouk E, Elachqar A, El Hallaoui A, et al. One-pot regioselective synthesis of N-benzoyl 2-amino-3,4-dihydro-3-oxo-2H-1,4-benzothiazines, Orient. J. Chem. 2010; 26 (4): 1249. Doi.org/10.3390/molecules15129354

[35] Mabrouk EH, Elachqar A, El Hallaoui A. Methyl 2-Benzamido-2-(1H-benzimidazol-1-ylmethoxy)acetate. Molbank. 2012; M777. Doi:10.3390/M777.

[36] Mabrouk EH, Elachqar A, El Hallaoui A. 2-[(1-Benzamido-2-methoxy-2-oxoethyl)amino]benzoic Acid. Molbank. 2013; M792. Doi:10.3390/M792.

[37] Mabrouk EH, Arrousse N, Korchi A, Lachgar M, Oubair A, Elachqar A, Jabha M, Lachkar M, El hajjaji F, Rais Z, Taleb M. Intelligence Way from Eco-friendly Synthesis Strategy of New Heterocyclic Pyrazolic Carboxylic α-Amino Esters. Chem. Res. Chinese Universities. 2020; 6: 1-7. Doi: 10.1007/s40242-020-0173-4.

Section 2

Furans and Furan Derivatives - Applications

Chapter 4

Furan Functionalized Polyesters and Polyurethanes for Thermally Reversible Reactive Hotmelt Adhesives

Laxmisha M. Sridhar, Andrew T. Slark and James A. Wilson

Abstract

New reactive hotmelt (RHM) adhesives based on thermally reversible Diels-Alder networks comprising multifunctional furan and maleimide prepolymers are described. The prepolymer mixture is easy to apply in the bulk from the melt and after application to the substrates, the adhesive undergoes polymerization at room temperature resulting in crosslinked bonds. Due to their thermoplastic nature and low melt viscosity at hot melt application temperatures, the adhesives provide processing properties similar to moisture cured polyurethanes (PUR). The technology is isocyanate-free and does not require moisture to initiate the crosslinking. Bonding and tensile properties of the RHM adhesive can be readily tuned by prepolymer design and provide cure rates similar to PUR adhesives. The Diels-Alder adhesives provide versatile adhesion to a variety of substrates and good creep resistance up to the retro temperature. The adhesives show good thermal stability during application and can be recycled multiple times by simple heating/cooling of the bonds providing similar performance. Several furan and maleimide prepolymers were scaled up to multi-Kg quantities to demonstrate the potential for industrial scalability. The results demonstrate that furan-maleimide reversible chemistry can be used for RHM application as a more sustainable alternative to conventional moisture curing PURs which tend to contain harmful residual isocyanate monomers.

Keywords: Furan, maleimide, polyurethane, thermally reversible, recyclable, moisture-free, isocyanate-free, de-bonding

1. Introduction

Several furan derivatives such as furfuryl alcohol, furfural and 5-hydroxymethylfurfural are bioderived. The industrial use of furan derivatives in polymeric applications is increasing - for example, furfuryl alcohol and furfural are used as raw materials to make furan resin, which is applied as a binder in polymeric concrete for construction [1]. 5-Hydroxymethylfurfural is a bioderived precursor for making 2,5-furandicarboxylic acid, which is a raw material used to make polyesters that are projected to replace phthalate plasticizers and even PET [2]. Other polymeric applications of furan derivatives include copolymerization in phenol-formaldehyde system [3], melamine-formaldehyde [4] and in biomedical applications [5]. In addition, benzofuran type highly conjugated furan functional polyaromatics have been used in organic solar cells [6]. On an industrial scale, Furnova Polymers Ltd. sells several furan based polymeric materials for concrete, anticorrosion, composites and high temperature applications. Another potential application of furan derivatives is reversible crosslinking to generate polymers that can be used to bond substrates and which can be reworked or de-bonded by depolymerization at high temperatures caused by the retro Diels-Alder reaction. Incorporation of furan-maleimide based thermally reversible covalent crosslinks into polymers is known [7]. Upon heating above 80°C, the crosslinks dissociate and upon cooling the crosslinks reform. This property has been used in a number of applications such as in interpenetrating networks [8], remendability [9] and self-healing [10]. Adhesives are widely used to bond substrates in a number of consumer and industrial applications [11]. Crosslinked adhesives provide the highest bond strength and durability but have a negative environmental impact, since permanent adhesion prevents easy separation at their end-of-life which does not facilitate the re-use or recycling of materials. With increasingly stringent requirements for meeting fuel economy and recycling targets, the automotive and electronic industry has a strong need for debondable/reworkable adhesives that enable efficient repositioning in defective assemblies or end-of-life cycle recycling. Adhesives have also been used to replace metal fasteners to make the assembly lighter and thus help meet the fuel economy target. There have been several reports on the use of maleimide-furan Diels-Alder networks for reversible adhesives that include epoxy [12], polyacrylates [13] and polyurethanes [14]. However, in these reports, a low molecular weight monomeric maleimide compound such as *N*,*N*′-(4,4′-methylene diphenyl) bismaleimide (BMI), which is readily available commercially, was used. However, this BMI is highly toxic and unsuitable for practical consumer or industrial adhesive applications. It also has poor compatibility with most resin systems because of its rigid aromatic backbone and high polarity. While Diels-Alder based reversible non-isocyanate polyurethanes have been described before [14], little is known on their material properties. Most of the reported reversible polymeric systems based on furan-maleimide Diels-Alder networks required exposure to hot solvents for effective removal of components. In summary, the incorporation of Diels-Alder chemistry in a truly practical fashion has not yet been demonstrated.

Traditional hot melt adhesives based on, for example, polyolefins or polyamides are thermoplastic polymers which melt upon heating so that they can be applied to a substrate in a fluid state when hot. The cohesive strength of these adhesives is derived primarily from intermolecular physical forces which are formed upon cooling the adhesive. Polyurethane RHM adhesives are also applied in molten form and initial (green) strength is provided on cooling into a solid state. However, these adhesives are reactive (made from isocyanate functional prepolymers) which cure by moisture over time at ambient temperature resulting in adhesives with improved performance, such as adhesion strength, heat resistance, toughness and chemical resistance [15] when compared to thermoplastic hotmelts. The isocyanate prepolymers are typically made by reacting excess diisocyanates such as 4,4′-methylenebisphenyl diisocyanate (MDI) with polyols. Practical use of this chemistry (lower viscosity at hotmelt application temperature) requires that a NCO/OH ratio in the range 1.5 to 2.3 should be used for the synthesis of isocyanate prepolymers. The resulting compositions usually contain residual diisocyanates such as MDI, which are respiratory and skin sensitizers. Despite the successful development of moisture curable RHM, the technology has some limitations. For example, RHMs must be stored in the absence of moisture to prevent premature curing and require specialized application equipment for processing to keep moisture out. The cure rates of RHM's can also vary depending on atmospheric humidity level, moisture content of the substrates, bond line thickness and moisture vapor transmission rate of the adhesive [16]. The compositions cure irreversibly and such adhesives cannot be used in applications where features such as repositionability, reworkability or recyclability are desired. There remains a need for improvement in RHM technology. Here we report on a new thermally reversible RHM that undergoes cross-linking at room temperature after application from the melt by copolymerization of polyfunctional furans and maleimides. Poly(ester urethane)-based prepolymers with maleimide end groups are developed and combined with the polyfunctional furans to provide thermally reversible polyurethane RHM's that are isocyanate-free and exhibit a range of physical properties similar to conventional polyurethane based RHMs. The new adhesives can be bonded, de-bonded and re-bonded multiple times without a significant change in performance.

2. Results and discussion

2.1 Initial proof-of-concept, ambient curing and cured adhesive strength

The thermally reversible RHM concept is schematically represented in **Figure 1** involving the dynamic equilibrium of Diels-Alder networks based on the combination of a polyfunctional furan crosslinker (indicated in blue color) and a maleimide prepolymer (indicated in red color). In the hotmelt state at higher temperatures (left hand side), network dissociation is favored - the Diels-Alder equilibrium is shifted toward free furan and maleimide functional prepolymers that provides the molten state. The fluid adhesive is then applied to substrates and after cooling, the prepolymers form crosslinked adhesive at ambient temperature via the [4 + 2] cycloaddition reaction between furan and maleimide functional groups (right hand side).

Reactive PUR hot melts are desirable due to their ease of application (involving low melt viscosity), high initial green (uncured) strength, curing at ambient temperature and versatile cured adhesion/durability. For initial investigations into the suitability of this concept, a variety of polyfunctional furans and maleimides were made in order to critically determine if they could behave like PUR hotmelts currently in use, i.e. (a) be applied as a fluid hotmelt at moderate temperatures, (b) crosslink at ambient temperature over a reasonable time period and (c) provide

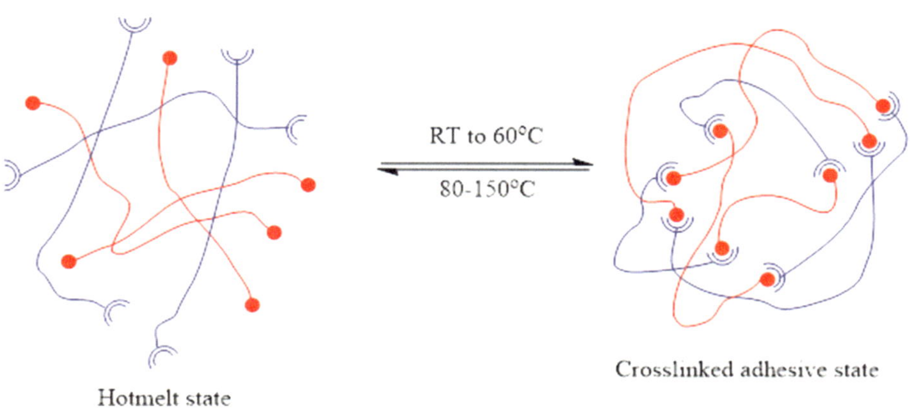

Figure 1.
Thermally reversible reactive hotmelt concept. For clarity, furan and maleimide functional reacting prepolymers are indicated in blue and red colors, respectively (see text for description).

suitable cured bond strengths. To this end, a number of polyfunctional furans and maleimides were made, formulated into reactive hot melts and compared to a reference commercially available PUR adhesive. Polyfunctional furans F-1, F-2, F-3, F-4, F-5 & F-6 comprising ester and urethane linkages were made (**Figure 2**). F-2, F-3, F-4 and F-5 were made by simple addition of bio renewable furfuryl alcohol to the corresponding commercially available isocyanates. To mimic polyurethane like properties, F-6 possessing a PU backbone was made by initially reacting polyester or polyether polyols with MDI followed by bulk addition of furfuryl alcohol in the same pot. Trifunctional furan F-1 was synthesized by reacting 1,3,5-benzenetricarbonyl chloride with furfuryl alcohol in the presence of triethylamine as base in a solvent [17]. When the isocyanate starting material are liquids, a bulk process can be used to make the corresponding polyfunctional furans. For example, F-2, F-5 and F-6 were made in the bulk without the use of solvents simply by slow addition of furfuryl alcohol to the corresponding isocyanates. The polyfunctional isocyanate starting material used for F-3 synthesis is commercially supplied in 25% butyl acetate solvent and it was used as supplied. Several batches of polyfunctional furans were made at multi-Kg scale to demonstrate the industrial scalability of the process [18].

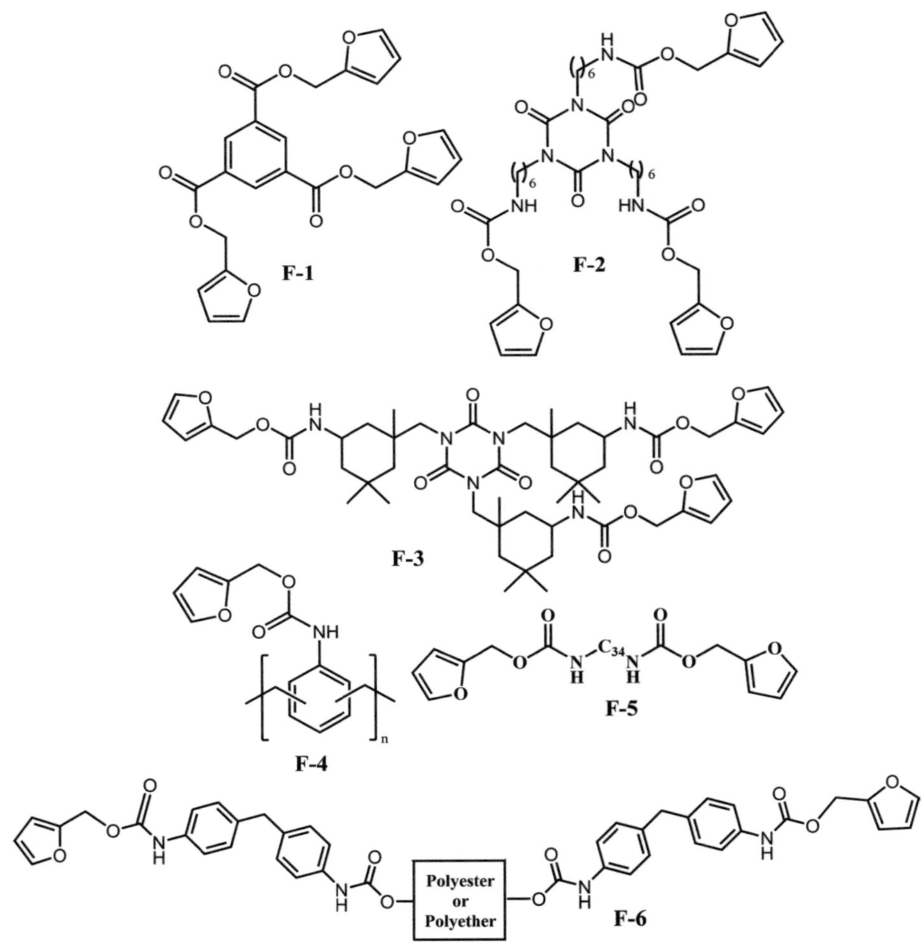

Figure 2.
Polyfunctional furan resins used in the study.

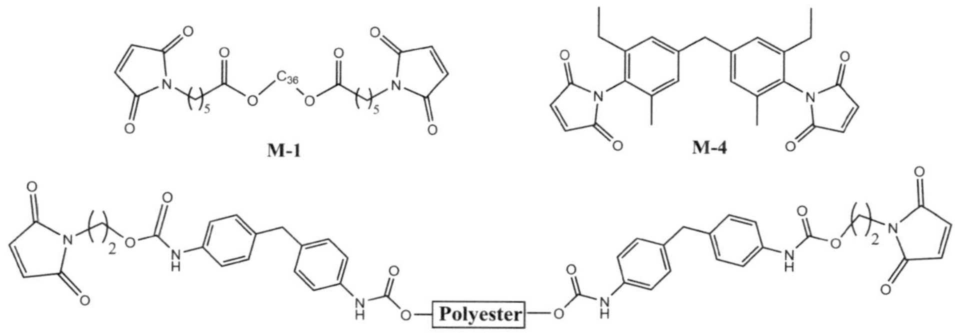

M-2, when polyester is Dynacol 7360

M-3, when polyester is Dynacol 7231

Figure 3.
Bismaleimide (BMI) resins used in the investigation.

Several bismaleimide resins (BMI) were also investigated as components in the study (**Figure 3**). M-1 is a liquid BMI supplied by Henkel Corporation and it has a flexible 36-carbon dimer diol backbone (represented as C_{36}). Two maleimide prepolymers M-2 & M-3, possessing poly(ester urethane) backbones were also used (discussed in detail in the later part of this chapter). Both the resins M-2 and M-3 were made in a bulk process by first reacting corresponding polyester polyols with excess MDI and subsequently capping the isocyanate terminated polyurethane prepolymers with 2-hydroxyethylmaleimide in a 2-step, one-pot process similar to the process used for the synthesis of F-6 [18]. The polyester polyols used for M-2 and M-3 (Dynacoll 7360 and 7231, respectively - obtained from Evonik) are used in typical moisture cured RHM formulations. Since excess diisocyanate is used in step 1 (1:2 stoichiometry of diol to diisocyanate used, some chain extension observed), a minor reaction product arising from the reaction of residual MDI remaining from step 1 with 2-hydroxyethylmaleimide was also observed. However, these adducts are nonvolatile as a consequence of their higher molecular weight and polarity (urethane & maleimide functionalities) as compared to MDI itself. Due to the presence of polyfunctional furans in the melt, they will be incorporated into the crosslinked Diels-Alder network. This is in stark contrast to conventional moisture cured RHM's based on isocyanate prepolymers, where the toxicity arises primarily from the volatility of residual MDI (at hotmelt application temperature) and potential hydrolysis to the corresponding aromatic amine. Prepolymers M-2 & M-3 possess semicrystalline and amorphous polyester backbones, respectively, and were used to investigate their impact on properties. A commercially available rigid aromatic bismaleimide M-4 (Daiwakasei industry) was also investigated as an additive in formulations to study its effect on properties. Several other polyfunctional maleimides were also made by simple addition of 2-hydroxyethylmaleimide to polyfunctional isocyanates by a process similar to that used for the synthesis of polyfunctional furans described in **Figure 2**. However, initial formulation work by blending these polyfunctional maleimides with furan prepolymers F-6 (possessing PU backbone) did not show much promise likely as a result of compatibility issues between highly polar polyfunctional maleimides and F-6. Hence, this approach was not pursued further.

Our initial formulation work focused on the feasibility of achieving room temperature cure and identifying structural features that contributed to bond strength at room temperature and 80°C. The components of formulas DA-1 to DA-7 shown in **Table 1** were mixed in the melt (typically at 120–150°C) until they became

Formula	F-1	F-2	F-3	F-4	F-5	M-1	M-2	M-3	M-4	Lap shear[a]	Lap shear[b]
DA-1	24.4					75.6				5.4	0.8
DA-2		36.8				63.2				5.8	0.7
DA-3			49.2			50.8				2.0	1.8
DA-4			17.5				25	57.5		5.0	0.9
DA-5				32		68				3.1	1.5
DA-6	27					63			10	3.3	1.8
DA-7	14				25	51			10	5.6	0.7
DA-8					44	56				n.d.	n.d.
Control[c] (PUR)										4.9	1.1

[a]*Lap shear strength tested at ambient temperature [beechwood substrate, 25 × 25 mm bond overlap], unit MPa.*
[b]*Lap shear strength tested at 80°C in air circulated oven [beechwood substrate, 25 × 25 mm bond overlap], unit MPa.*
[c]*A commercial isocyanate prepolymer from Henkel corporation was used as a reference.*
n.d. means not determined.
The lap shear samples were tested using a JJ Lloyd tensiometer, with a load cell of 10 kN & a crosshead speed of 100 mm.min^{-1}.

Table 1.
Diels-Alder formulations investigated in the study. The formulation components shown are in weight % (wt%).

completely homogenous followed by coating on Beechwood substrate (250 μm coating thickness, see **Table 1**) and subsequent bonding.

The bonded substrates were allowed to cure at 23°C for 1 week at 50% relative humidity (RH) before testing for lap shear strength. As a control, a commercial isocyanate terminated polyurethane prepolymer (PUR) from Henkel Corporation was used. This control was bonded, cured at room temperature for 1 week and tested in the same way as the Diels-Alder based formulations.

Formula DA-1, which used a flexible BMI resin M-1 showed good curability at room temperature as evidenced by good lap shear strength development (DA-1 v PUR). Owing to the inherent thermal reversibility of the network, the lap shear strength at 80°C was slightly inferior to the control PUR. The addition of 10 wt% of rigid aromatic BMI M-4 improved the hot strength at 80°C (DA-6 vs. DA-1). However, the room temperature curability of DA-6 was negatively affected as evidenced by the lower lap shear strength. This is likely a result of a higher glass transition temperature (T_g) of the network which would restrict mobility of the reactants at room temperature. By partially substituting F-1 with flexible bifunctional F-5 (maintaining crosslinks in the network), the room temperature curability was improved at the expense of 80°C strength (DA-7 v DA-6). Similar observations were made with formulas DA-2 and DA-3. Use of flexible furan F-2 in DA-2 gave good lap shear strength at room temperature. In contrast, rigid furan F-3 used in formula DA-3 improved the hot lap shear strength at the expense of room temperature strength. Formula DA-5 that contains F-4 with a rigid aromatic backbone showed lap shear performance similar to DA-3 at both room temperature and 80°C. For comparison, a linear Diels-Alder network (DA-8 formula) was made by blending bifunctional precursors F-5 and M-1 and allowing to copolymerize at room temperature for 3 days. A size exclusion chromatography (SEC) molecular weight comparison of this polymer with M-1 and F-5 showed a several-fold increase in molecular weight arising from step-growth polymerization at room temperature (**Figure 4**). In addition, there were very low levels of reactants M-1 & F-5 present. While both M-1 and F-5 are medium viscosity liquids at ambient temperature (2,000–5,000 cP), the linear polymer was a gel like material at room temperature

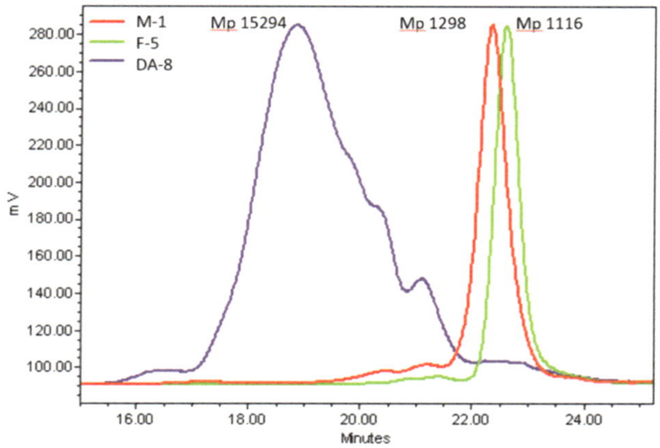

Figure 4.
*SEC chromatogram comparison for M-1 (M_p = 1298 Da), F-5 (M_p = 1116 Da) and the corresponding linear Diels-Alder polymer DA-8 (M_p = 15294 Da). Mp refers to peak molecular weight. The chromatograms were run using Waters e2695 instrument comprising three Agilent Polypore 7.8*300 mm GPC columns using THF as mobile phase and Waters 2414 as a RI detector against polystyrene standards.*

and a liquid at 60°C as a likely consequence of low T_g (both M-1 and F-5 possess flexible backbones). No lap shear measurements were performed on this system. The molecular weight for the polymer is likely to increase further if left at room temperature for a longer time period since it would take about a week for high prepolymer conversion (discussed later in the chapter). The SEC data presented here further demonstrates the room temperature reactivity of the Diels-Alder based systems.

The cause of contrasting lap shear results obtained with formulations DA-2 and DA-3, which contain the same maleimide component M-1 but structurally different furans F-2 and F-3, respectively, was investigated by differential scanning calorimetry (DSC) studies. It should be noted that F-2 contains a flexible backbone while F-3 possesses a highly rigid cycloaliphatic backbone. **Figure 5** shows DSC thermograms for neat F-2 and F-3, and the corresponding formulas DA-2 and DA-3, respectively. Polyfunctional furan F-2 appears to be completely compatible in the DA-2 formulation with only a T_g at 10°C for the crosslinked network and

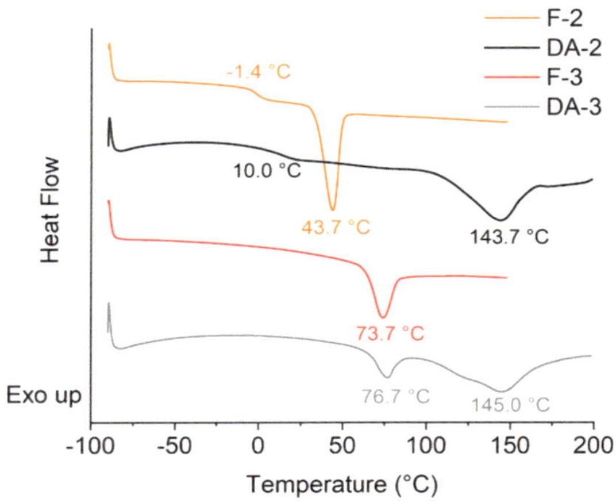

Figure 5.
DSC thermograms (TA instruments Discovery DSC 25) of F-2, DA-2, F-3 and DA-3. Reproduced from [17] with permission from Royal Society of Chemistry.

an endotherm above 100°C corresponding to the retro Diels-Alder depolymerization. No residual melting point for F-2 was recorded. In contrast, the cured DA-3 network shows a residual melting point at 76.7° C for F-3 along with the expected endotherm for the depolymerization. This indicates that the rigid F-3 is partially incompatible at room temperature in DA-3, which could explain the lower lap shear strength obtained at room temperature (DA-3, **Table 1**). The partial incompatibility could also be resulting from the higher concentration of F-3 (higher molecular weight) required in DA-3 (as compared to DA-2) for 1:1 molar equivalence between furan and maleimide functionalities. The higher T_g of the partially cured network is also likely contributing to incomplete polymerization at room temperature. The DA-3 formula likely needs to be heated to above the melting point of F-3 for better compatibility, which is reflected in higher lap shear strength observed at 80°C. A similar effect may be operating in DA-5, which uses rigid furan F-4, although this was not investigated in detail.

Since the Diels-Alder network formation does not require moisture for curing, in principle the cure rate is independent of moisture availability. **Figure 6** shows comparison of cured films of DA-1 and the moisture cured PUR control. The DA-1 cured film appears clean without bubbles (**Figure 6**, left image). In contrast, the moisture curing PUR control shows significant bubble generation arising from CO_2 evolution during moisture cure (right image). If CO_2 cannot diffuse out of the bond line during application, the pressure generated can cause substrate deformation, especially when non-porous substrates are bonded together, such as plastics and metals.

Overall, these preliminary results indicate that a combination of both soft and rigid backbones are required in the prepolymers to obtain a balance of adhesion strength at room temperature and 80°C. The effect of T_g of the crosslinked network on prepolymer conversion at room temperature is discussed in detail at a later section. Formula DA-4, which used the rigid F-3 crosslinker and two bismaleimides M-2 and M-3 possessing poly(ester urethane) backbones showed lap shear performance very similar to the PUR control. This initial investigation showed that the new thermally reversible RHM could be applied as fluid hotmelts at moderate temperatures, would cure at similar rates to conventional PUR at ambient temperature and that similar bond strengths could be achieved. Based on these initial results, DA-1 and DA-4 were selected for more detailed investigations.

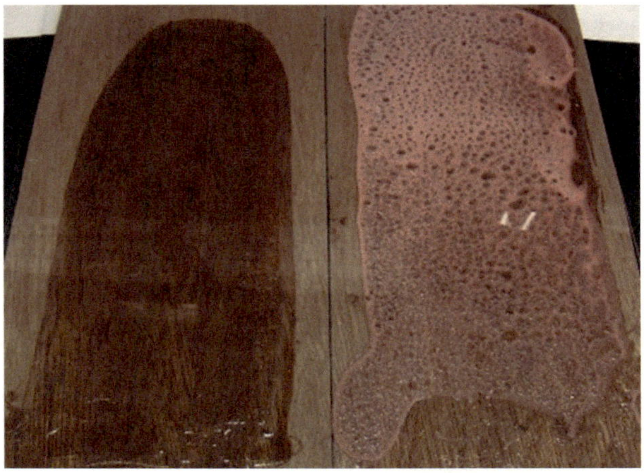

Figure 6.
Appearance of cured films of DA-1 (left) and moisture cured PUR control (right).

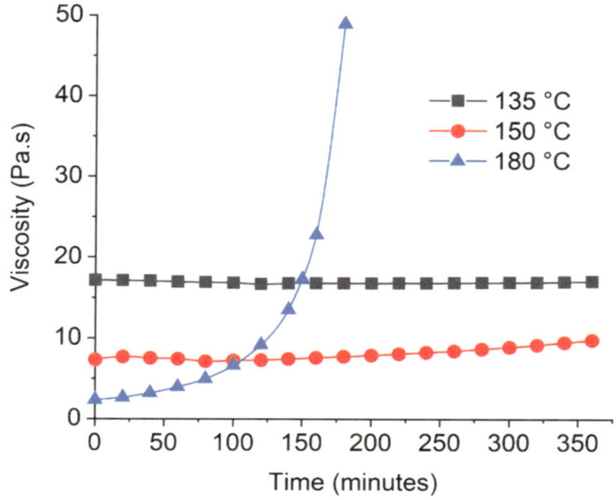

Figure 7.
Melt viscosity of DA-4 at different application temperatures as a function of time (run using Brookfield Viscometer model RVDV-1 with a Model 74R temperature controller and Thermosel unit using spindle no. 27). Reproduced from [17] with permission from Royal Society of Chemistry.

2.2 Thermal stability

In real world applications, RHM adhesives are held at the desired application temperature (e.g. in the range 120–150°C) for a number of hours until all the substrates are coated and bonded in the production line. This requires good thermal stability up to several hours in the open air. To investigate the stability of the Diels-Alder RHM system, DA-4 (which is compositionally closest to PUR) was held at a specified melt temperature and the viscosity was measured as a function of time using a Brookfield viscometer (**Figure** 7). The initial melt viscosity is in the range 7,000–17,000 mPa.s at 135°C and 150°C, similar to the melt viscosity of PUR products which provide good substrate wetting. The DA-4 formula also shows good thermal stability at 135°C and 150°C up to about 6 hours. It should be noted that the viscosity profile of the DA-4 at these temperatures is very similar to the moisture cure PUR benchmark. To test the limit of thermal stability, DA-4 was also held at 180°C for several hours. However, significant viscosity increase was observed likely arising from homopolymerization of the free maleimide component. It should be noted that this is not a typical temperature used for reactive hotmelts in current industrial applications since the NCO prepolymers would be highly unstable at these temperatures as trimerization of isocyanate functionality is possible.

2.3 Green (initial) strength

An important processing property requirement of RHM adhesives is green strength, which is the initial bond strength soon after the adhesive is applied from the melt and before moisture curing takes place. Liquid NCO prepolymers applied at ambient temperature have very little green strength as there is little cohesion to keep substrates apart, and substrates need to be clamped together before sufficient strength is built up. However, reactive PUR hot melts have the advantage that the green strength is relatively high and additional steps are not required to hold substrates together before curing takes place. This results from non-covalent interactions providing physical strength on cooling, provided by hydrogen bonding and crystallization. DA-1 and DA-6 (similar formulas but DA-6 with 10% additive) were

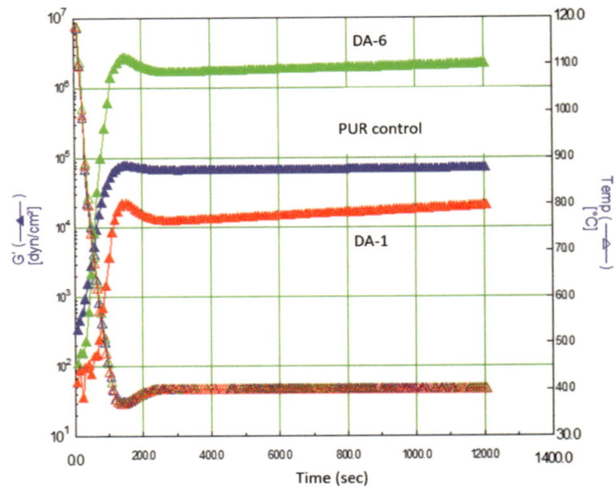

Figure 8.
Green strength development comparison of DA-1 (red), DA-6 (green) with PUR control (blue). Test performed on ARES-M rheometer (TA instruments) using dynamic oscillatory test method in which 10 rad/s oscillating frequency was used with 1% strain and the data recorded at 11 second intervals.

evaluated for green strength in comparison to the PUR reference by rheological analysis. Experiments were performed with rapid cooling where adhesive compositions were positioned between parallel plates cooled from application temperature to 40°C at a cooling rate of 60°C per minute in nitrogen atmosphere, and then held for 20 minutes to evaluate the storage modulus (elastic component) as a function of time. **Figure 8** shows that DA-1 (comprising a very flexible aliphatic backbone) has a lower modulus than the PUR reference. However, this can be tuned since DA-6 (similar to DA-1 but with 10% additive) provides higher green strength than the reference as a consequence of the incorporation of a rigid, aromatic backbone. These results indicate that the green strength of Diels-Alder based RHM systems can be tuned by the appropriate choice of prepolymer composition.

2.4 Cure dynamics and adhesion development

FT IR spectroscopy was used to investigate the ambient temperature conversion of the DA-1 composition after deposition from the melt by monitoring the characteristic maleimide absorption at 696 cm^{-1} (**Figure 9**). In the cured film (top IR trace indicated in black color), the 696 cm^{-1} band is not visible. After heating the film at 150°C for 1 h and immediately cooling, this band is prominent initially but decreases in intensity as the crosslinking progresses. The decrease is rapid in the first two days after deposition followed by a slower change in the next several days. The band is barely visible after 10 days of storage at room temperature.

In a separate experiment, the composition DA-1 was applied from the bulk between two Beechwood substrates (25 × 25 mm bond overlap) and bond strength development was compared over time at room temperature and at 60°C (**Figure 10**). At room temperature, the bond strength increases substantially over a period of 48 hours before tending to a plateau. This is consistent with the decrease of the maleimide absorption peak at 696 cm^{-1} as Diels-Alder cycloaddition progresses as shown in **Figure 9**. It is interesting to note that substrate failure begins to occur at a bond strength of about 4 MPa and that this is reached in approximately 24 hours at room temperature. In comparison, the strength development at 60°C is more rapid and reaches 4 MPa in approximately 2 hours.

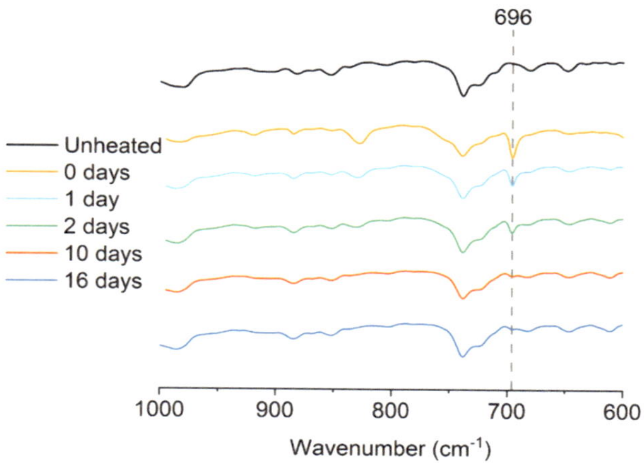

Figure 9.
IR spectra of DA-1 cured network over the range 600–1000 cm^{-1} before heating (top) and over time at ambient temperature after deposition from the melt. IR measurements were recorded using a PerkinElmer FT-Infra-red spectrometer equipped with a UATR Two accessory. Reproduced from [17] with permission from Royal Society of Chemistry.

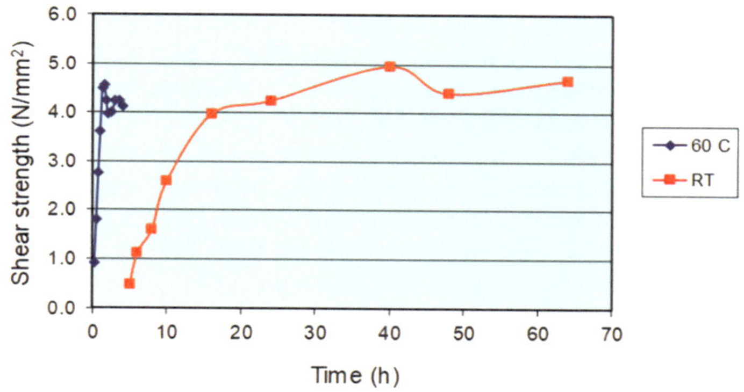

Figure 10.
Comparison of lap shear strength development on Beechwood substrate for DA-1 formulation at room temperature (RT) and 60°C as a function of time.

2.5 Cured properties and adhesion

For isocyanate terminated PUR moisture cure adhesives, the mechanical performance is determined by a combination of covalent crosslinking formed by moisture curing (including urea formation), H-bonding of urethane groups and crystallinity from polyester segments present in the prepolymer backbone. The urethane and urea groups contribute to the flexibility & elongation of the cured network while the crystallinity of the polyester segment causes further reinforcement of modulus and stiffness. The results discussed previously demonstrated that the lap shear strength, room temperature curability and mechanical properties can be tuned by proper choice of polyfunctional furans possessing rigid or flexible backbones. However, to achieve properties similar to moisture cured PUR, the use of polyurethane backbones with reinforcing crystalline segments is necessary. We sought to achieve this initially by making furan functional PU prepolymers F-6 possessing semicrystalline polyester segments and blending them with polyfunctional maleimides obtained by the addition of 2-hydroxyethylmaleimide with polyfunctional isocyanates. However, our initial results did not show promise likely as a result of compatibility issues. As an alternative approach, maleimide terminated PU prepolymers M-2 & M-3 possessing poly(ester urethane) backbones were made (**Figure 3**) in a one-pot bulk process and formulated with polyfunctional furans. The synthetic process is very similar to that utilized for the synthesis of isocyanate terminated PUR prepolymers except that an additional solvent-free process was used to react the terminal isocyanate groups with 2-hydroxyethylmaleimide in the same-pot [18]. The prepolymers were scaled up to several Kgs with reproducible performance (discussed in the next section) to demonstrate the industrial scalability of the process. A blend of M-2 comprising a semi-crystalline backbone and M-3 with amorphous polyester segment was used to achieve a balance of room temperature and 80°C lap shear strength. Such combinations of semi-crystalline and amorphous polyester polyol segments are also used in moisture cured PUR products to provide the necessary balance of properties suitable for applications.

The DSC thermogram for the cured DA-4 network shows a T_g of −11.7°C and a melting point (T_m) at 50°C for the semi-crystalline polyester segment present in copolymerized prepolymer M-2 (also seen in the DSC thermogram of M-2). In addition, there is an endotherm at 140.4°C resulting from the retro Diels-Alder depolymerization (**Figure 11**). We have previously shown that the position of this endotherm is a function of the heating rate, with much lower temperatures obtained at 1°C per minute compared to 10°C per minute [17]. Compared to DA-4, the thermograms for DA-2 and DA-1 do not exhibit a melting point due to the amorphous backbones but have higher T_g values of 10°C and 47.3°C, respectively. In addition, the peak position and magnitude of the endotherms due to the retro Diels-Alder reactions are relatively low for DA-4 compared to the DA-1 and DA-2. This is due to the relatively lower crosslink density in DA-4 arising from lower concentration of furan and maleimide functional groups since the molecular weights of M-2 and M-3 are much higher compared to M-1 used in DA-1 and DA-2. A similar phenomenon has been observed in self-healing coatings [19]. It is important to note that no residual melting point for F-3 was seen in DA-4, which indicates good compatibility in the formula (compared to DA-3, **Figure 5**).

The conversion in the Diels-Alder polymerization of DA-1, DA-2 and DA-4 formulas was investigated after deposition from the melt by monitoring the characteristic absorption at 696 cm^{-1} for the maleimide functionality. **Figure 12** shows the relative intensities of this band as a function of time. The initial absorption

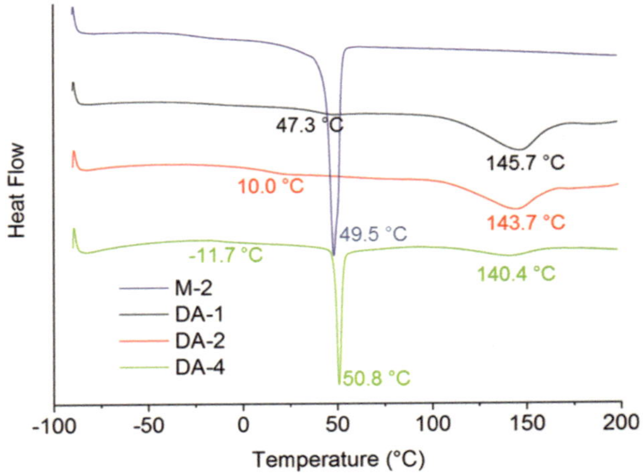

Figure 11.
DSC thermogram for DA-4 and comparison with M-2, DA-1 and DA-2 (TA instruments Discovery DSC 25). Reproduced from [17] with permission from Royal Society of Chemistry.

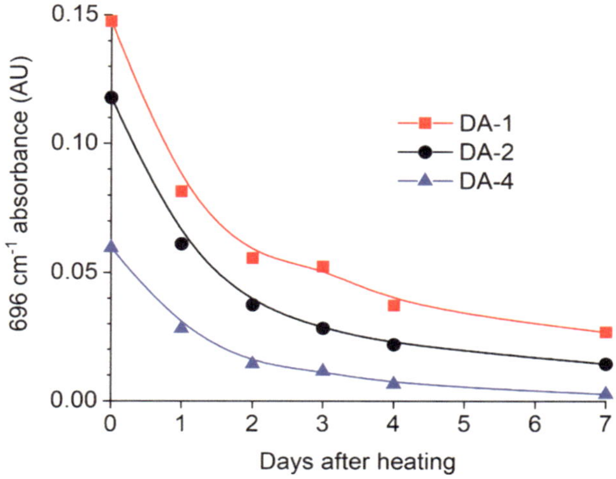

Figure 12.
IR absorbance comparison of maleimide absorption at 696 cm^{-1} as a function of time for DA-1, DA-2 and DA-4. Absorbance values after 7 days are: 0.027 for DA-1 (82% conversion), 0.014 for DA-2 (88% conversion) and 0.003 for DA-4 (95% conversion). Reproduced from [17] with permission from Royal Society of Chemistry.

intensities correspond well to the relative concentration of the maleimide functionality in the 3 formulas (DA-1 > DA-2 > DA-4). In spite of the significant difference in reactive functional group density, prepolymer molecular weight and backbone, all the three formulas showed a similar reactivity tendency in the first 48 hours. There was fast reduction in the first 48 hours, followed by slower decrease, before reaching a plateau at 7 days. A closer look at the relative absorption after 7 days however showed some difference in conversion. DA-4 with the lowest T_g (−11.7°C) shows the highest conversion (95%) while DA-1 with the highest T_g shows the lowest conversion (82%). This result is consistent with higher mobility typically seen in lower T_g networks. The lower room temperature lap shear strength previously observed with compositions DA-3, DA-5 and DA-6, which used polyfunctional furans with rigid backbones, are also likely to result from lower conversion due to restricted mobility (**Table 1**). These formulas showed better 80°C lap shear strength than the PUR reference.

The bond strengths of DA-1 & DA-4 were compared to the control PUR system on three different types of material substrates – wood, plastic and metal (**Figure 13**). For bonding Beechwood substrates at ambient temperature, the lap shear strengths of both Diels-Alder networks were similar to the reference PUR. However, at 80°C the bonding performance was slightly better for the irreversible PUR system (**Figure 13A**). The bond strength of DA-4 on polycarbonate (PC) was slightly superior to the reference PUR at ambient temperature (**Figure 13B**). On aluminum substrate in particular, the bond strength was exceptional for the Diels-Alder formula. These results demonstrate the versatile adhesion performance of the Diels-Alder based networks using different substrates and surface chemistry.

2.6 Cured mechanical properties

The mechanical properties of thermally reversible DA-1 and DA-4 were compared to the moisture cured PUR benchmark. As a result of higher crosslink density obtained from low molecular weight prepolymers, DA-1 exhibits the highest break stress and lowest elongation (**Table 2**). Due to the higher molecular weight backbones used in M-2 and M-3 prepolymers, DA-4 has a lower crosslink density. This results in stress–strain and elongation properties similar to PUR benchmark.

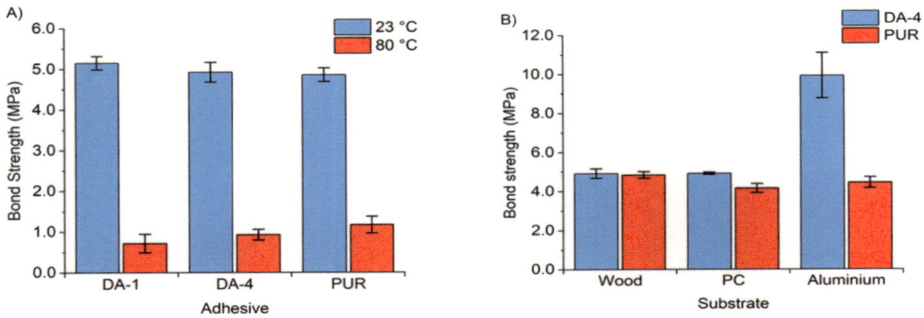

Figure 13.
(A) Lap shear strength comparison of DA-1, DA-4 and control PUR at ambient temperature and 80°C using Beechwood substrate. (B) Bond strength comparison of DA-4 and control PUR using different substrates at ambient temperature. Reproduced from [17] with permission from Royal Society of Chemistry.

Formula	Modulus[a]	Break Stress[a] (MPa)	Elongation at break[a] (%)	Creep failure temperature[b] (°C)
DA-1	70	10.5	140	90
DA-4	124	7.7	480	80
Control PUR	93	5.4	735	>110

[a]*Determined by tensiometric analysis (100 N load cell, 100 mm.min*$^{-1}$*).*
[b]*Determined by vertical creep measurements (25 × 25 mm bond overlap, 1 Kg static load.*

Table 2.
Mechanical properties of films and creep resistance measurements for DA-1 and DA-4 compared to PUR control.

It should be noted that DA-4 and the reference PUR both comprise semi-crystalline polyester segments at a similar overall concentration.

The mechanical properties of DA-1, DA-2 and DA-4 Diels-Alder networks were evaluated by DMTA using 250 μm cured films (**Figure 14**). The 3 networks have high stiffness below their respective T_g. DA-4 has the highest mechanical properties due to the presence of the semi-crystalline segments resulting from M-2 and the modulus changes significantly above ambient temperature as a consequence of melting (**Figure 14A**). Above the T_g and T_m (for DA-4), the change in modulus is slow for the 3 networks in the temperature range 40–80°C. There is then rapid

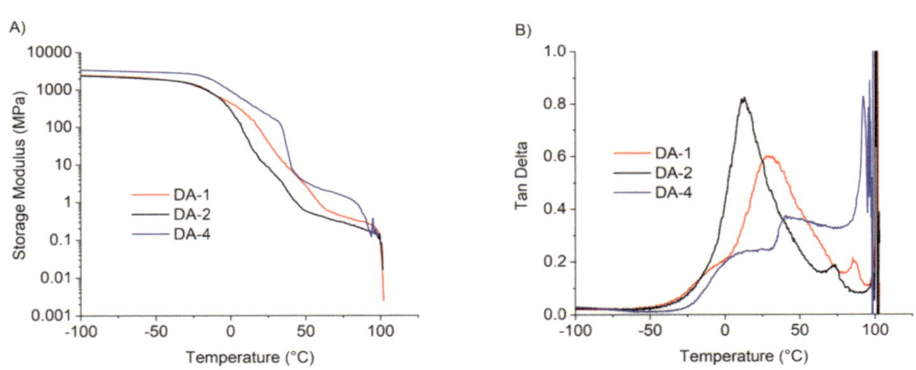

Figure 14.
(A) Storage modulus vs. temperature of DA-1, DA-2 and DA-4. (B) Tanδ vs. temperature of DA-1, DA-2 and DA-4 from DMTA on 250 μm films. Reproduced from [17] with permission from Royal Society of Chemistry.

decrease in modulus at higher temperature (above 80°C) with the tan δ reaching 1 (**Figure 14B**), at which point there is a transition from a solid to a liquid phase. This is the temperature range where the Diels-Alder equilibrium shifts predominately to the prepolymeric dissociated state resulting in stretching and failure of the film samples being tested.

Comparison of crosslink densities of the Diels-Alder networks shows a correlation between crosslink density and the rapid decrease in storage modulus (*G*') above 80°C. DA-1 has a higher functional group concentration because of lower molecular weight reactants (1.63 mol.Kg^{-1}, calculated based on stoichiometry of reagents), which results in higher crosslink density. Even though the Diels-Alder equilibrium starts to shift to the prepolymeric state above 80°C, the higher crosslink density present in DA-1 delays the rapid decrease in *G*' to approximately 97°C. In contrast, DA-4 comprising a lower crosslink density (lower functional group concentration of 0.39 mol.Kg^{-1}) shows a rapid change in modulus at approximately 85°C.

A further important property of RHM adhesives is their creep behavior at higher temperature under static load. The creep resistance of reversible DA-1 and DA-4 was compared with the irreversible moisture cure PUR control up to 110°C (**Table 2**). This upper temperature limit was chosen based on tan δ data, which shows liquid-like behavior for DA-1 & DA-4 networks (**Figure 14**). Creep experiments were performed on bonded joints under a static load of 1 Kg in an oven held at a particular temperature for 24 h. The creep resistance of the reversible Diels-Alder system correlates well with the crosslink density of the cured network. The DA-1 system with a higher crosslink density was resistant to creep at 80°C but the joints failed at 90°C. The DA-4 system which has a lower crosslink density failed at 80°C but was resilient to creep at 70°C. The irreversible PUR control was creep resistant over the entire temperature range. Both the Diels-Alder networks were creep resistant in a temperature range well above their T_g and T_m (for DA-4). This result indicates a significant advantage of these reversible Diels-Alder based networks as compared to some of the reported creep behavior of other covalently adaptable networks [20]. The failure temperature observed in the creep experiment of DA-1 and DA-4 networks correlates well with their respective DMTA data, which indicates significant change in stiffness close to the failure temperature (**Figure 14**). Since the data shows strong dependency of creep resistance on crosslink density, it is feasible to develop new Diels-Alder systems with improved performance by changing the architecture of the furan and maleimide components.

2.7 Reversibility and recyclability

For a truly sustainable adhesive, its material properties and adhesion should not change significantly after multiple reprocessing and re-use cycles. In several industrial applications (including assembly of components in the automotive industry) the bonded substrates need to be repositioned if a defect is found or recycled at the end of their life. This requires that the adhesive shows similar performance when subjected to multiple bond, de-bond, re-bond cycles. The repeatability of lap shear adhesion of DA-1 and DA-4 formulas over multiple cycles was tested using adhesive films cured between aluminum joints. Initial bonds were formed using a film adhesive that was heated to the hotmelt dissociated state to wet the substrates and then cured for 7 days at ambient temperature before bond strength measurement. For the 2nd, 3rd and 4th re-use, the same joints were broken after the lap shear test and reassembled using the residual adhesive on the surface by heating and re-bonding of the broken joints. The adhesion data over multiple cycles is shown in **Figure 15**. Both DA-1 and DA-4 are relatively robust in performance after multiple bond, de-bond,

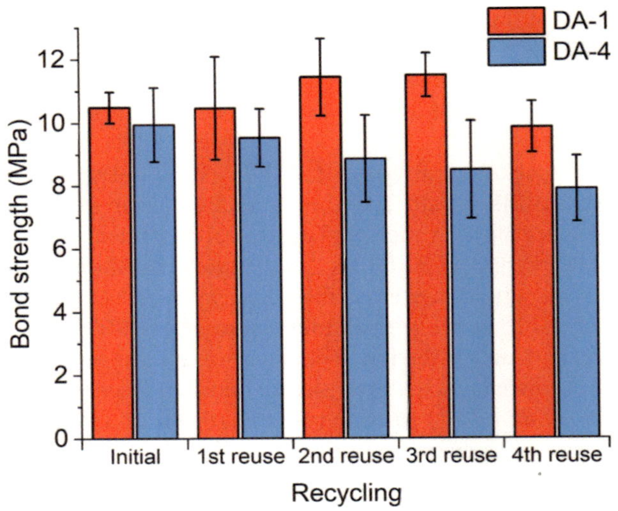

Figure 15.
Bond strengths of DA-1 and DA-4 over several cycles of bonding, de-bonding and re-bonding using the same aluminum substrate. Reproduced from [17] with permission from Royal Society of Chemistry.

re-bond cycles. The bond strength of the DA-1 system is unchanged after 4 re-use cycles while DA-4 shows a small decrease in strength (**Figure 15**). This is a significant result despite the likelihood that some adhesive could be lost or migrate out of the bond line during multiple heating, re-bonding and lap shear testing.

2.8 Development of the concept to soft cured Diels-Alder networks

As an extension of the RHM study, an initial investigation was performed on Diels-Alder based systems that give cured networks with a lower modulus at ambient temperature, approaching the pressure sensitive adhesive (PSA) region (10^4 to 10^5 Pa). This study required the synthesis of higher molecular weight maleimide prepolymer M-5 comprising an amorphous polyester segment with a lower T_g. To this end, M-5, containing Dynacol 7250 backbone (molecular weight 5500 g. mol^{-1} and T_g – 50°C as compared to 3500 g.mol^{-1} and T_g – 30°C for Dynacol 7231 used previously), was made using a process similar to that used for M-3 synthesis (**Figure 16**) [18]. An additional furan F-7 possessing a very soft backbone was also made by simple addition of furfuryl alcohol to Desmodur XP2599 (Covestro). The structure of this commercially available isocyanate (functionality >3) is not known and the stoichiometry for the synthesis of F-7 was calculated based on isocyanate equivalent weight.

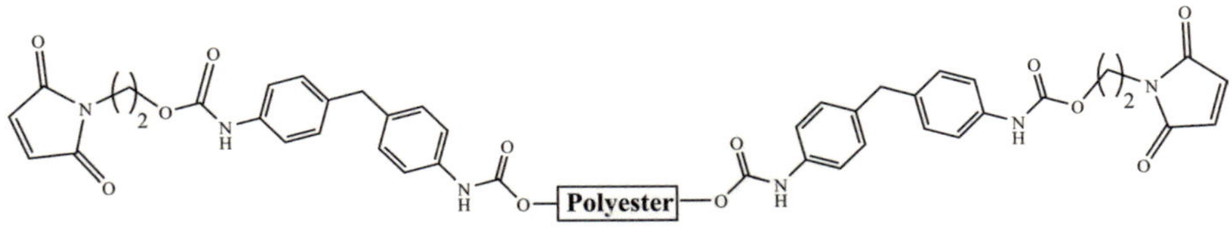

M-5, when polyester is Dynacol 7250

F-7 (polyfunctional furan derived from Desmodur XP2599)

Figure 16.
Prepolymers used for the development of soft Diels-Alder cured networks.

Formula	F-2	F-7	M-5	Modulus (Pa)[a]	Elongation at break (%)[a]
DA-9	8		92	6.2×10^5	658
DA-10		20.2	79.8	5.1×10^5	579

[a]*Determined by tensiometric analysis (100 N load cell, 100 mm.min^{-1}).*

Table 3.
Diels-Alder formulas used for the development of soft cured networks. The formulation components shown are in weight %.

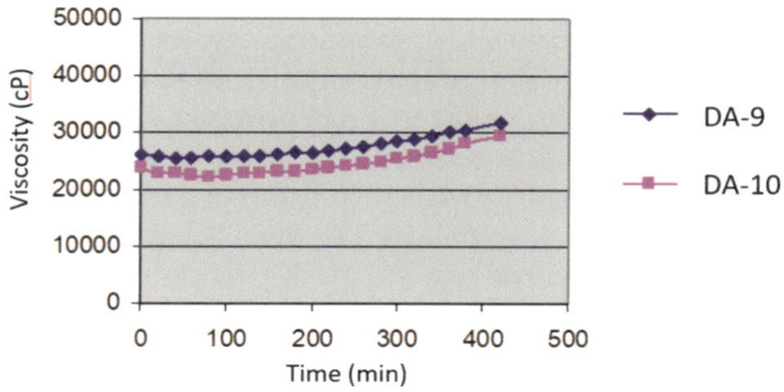

Figure 17.
Melt viscosity stability of DA-9 and DA-10 formulas as a function of time at 135°C.

The cured network DA-9 comprising F-2 and M-5 showed high elongation and a low modulus. DA-10 comprising F-7 and M-5 showed even lower modulus and similar elongation properties (**Table 3**). **Figure 17** shows that melt viscosity remains relatively low for these types of formulation and that thermal stability is good over several hours at 135°C. While these amorphous networks with relatively low crosslink density are not expected to have the same level of creep resistance as the networks DA-1 and DA-4 required for RHM applications, they may be sufficient for PSAs. In conventional solvent borne PSA technology, permanent crosslinking is typically achieved by the use of low level of metal chelates, which upon evaporation of the solvent bind with pendant hydroxyl or carboxylic functionality from the polymer backbone and provide crosslinks [21]. The degree of crosslinking is controlled by the amount of metal chelate used in the formulations. The crosslinking mediated by the Diels-Alder reaction could potentially be an effective metal-free substitute for new PSA development especially as a hotmelt PSA. In hotmelt PSA's, the adhesive is applied from the melt to substrates (no solvent used unlike solvent borne PSA and environmentally friendly) and the adhesive after application to substrates requires a crosslinking mechanism (such as UV induced crosslinking) for sufficient shear adhesion strength. Since the depolymerization gives the melt state (and sufficiently low viscosity to wet the substrates), the Diels-Alder mediated room temperature crosslinking could be an effective alternative method to achieve crosslinking in hotmelt PSAs. More formulation work (for example with tackifiers) is needed to develop formulations that would meet the tack, peel and shear adhesion requirements.

3. Conclusions

Polyester and polyurethane functional furans were reversibly copolymerized in the bulk with multifunctional maleimide prepolymers in a successful demonstration of a thermally reversible reactive hotmelt (RHM) adhesive concept that is isocyanate-free. At the hotmelt temperature, the adhesive remained in the pre-polymeric dissociated state with a low melt viscosity, enabling facile application to substrates and bond formation. Upon cooling, ambient temperature cure took place mediated by the Diels-Alder cycloaddition reaction. The reversible Diels-Alder RHM adhesive showed good thermal stability at the hotmelt application temperature. The green strength obtained after immediate cooling was highly tunable by choice of the prepolymers used in the formulation. The cure rate observed with the reversible RHM adhesive was similar to the moisture cure PUR benchmark and the new RHM adhesive showed versatile adhesion on a range of substrates. Mechanical and tensile properties were tunable depending on the choice of backbone present on the prepolymers used. A strong correlation between crosslink density and creep performance was observed. Repeated bonding, de-bonding and re-bonding experiments demonstrated a similar level of performance over multiple cycles. The adhesive reversibility would enable recycling or repositioning of bonded components several times without significant deterioration in performance. Diels-Alder networks with lower crosslink density have also been developed, which could be adapted for hotmelt pressure sensitive adhesive applications. Several polyfunctional furans and maleimide prepolymers were synthesized in multi-Kg scale to demonstrate industrial scalability.

Acknowledgements

The authors wish to thank Donald E. Herr for useful discussion.

Conflict of interest

The authors declare no conflicts of interest.

Author details

Laxmisha M. Sridhar[1*], Andrew T. Slark[2] and James A. Wilson[2]

1 Henkel Corporation, Rocky Hill, CT, USA

2 Department of Chemistry, University of Sheffield, Brook Hill, Sheffield, UK

*Address all correspondence to: laxmisha.sridhar@henkel.com

References

[1] Muthukumar M, Mohan D. Studies on furan polymer concrete. Journal of Polymer Research. 2005;12:231-241. DOI: 10.1007/s10965-004-3206-7

[2] Guidotti G, Soccio M, Garcia-Gutierrez MC, Ezquerra T, Siracusa V, Gutierrez-Fernandez E, Munari A, Lotti N. Fully biobased superpolymers of 2,5-furandicarboxylic acid with different functional properties. ACS Sustainable Chemistry & Engineering. 2020;8:9558-9568. DOI: 10.1021/acssuschemeng.0c02840

[3] Abdulla S, Pizzi A, Bahabri F, Ganash A. Maderas. Furanic copolymers with synthetic and natural phenolic materials for wood adhesives-A MALDI TOF study. Ciencia y Tecnologia. 2015;17: 99-104. DOI:10.4067/S0718-221X2015005000010

[4] Urdl K, Weiss S, Karpa A, Peric M, Zikulnig-Rusch E, Brecht M, Kandelbauer A, Muller U, Kern W. Furan-functionalized melamine-formaldehyde particles performing Diels-Alder reactions. European Polymer Journal. 2018;108:225-234. DOI: 10.1016/j.eurpolymj.2018.08.023

[5] Gevrek TN, Sanyal A. Furan-containing polymeric materials: Harnessing the Diels-Alder chemistry for biomedical applications. European Polymer Journal. 2021;153:110514. DOI: 10.1016/j.eurpolymj.2021.110514

[6] Islam A, Liu ZY, Peng RX, Jiang WG, Lei T, Li, W, Zhang L, Yang, RJ, Guan Q, Ge ZY. Furan containing conjugated polymers for organic solar cells. Chinese Journal of Polymer Science. 2017;35:171-183. DOI: 10.1007/s10118-017-1886-9

[7] Gandhini A. The furan/maleimide Diels–Alder reaction: A versatile click–unclick tool in macromolecular synthesis. Progress in Polymer Science. 2013;38:1-29. DOI: 10.1016/j.progpolymsci.2012.04.002

[8] Imai Y, Itoh H, Naka K, Chujo Y. Thermally reversible IPN organic–inorganic polymer hybrids utilizing the Diels–Alder reaction. Macromolecules. 2000;33:4343-4346. DOI : 10.1021/ma991899b

[9] Chen X, Wudl F, Mal AK, Shen H, Nutt SR. New thermally remendable highly cross-linked polymeric materials. Macromolecules. 2003;36:1802-1807. DOI: 10.1021/ma0210675

[10] Zeng C, Seino H, Ren J, Hatanaka K, Yoshie N. Bio based furan polymers with self-healing ability. Macromolecules. 2013;46:1794-1802. DOI: 10.1021/ma3023603

[11] Ozer H. Applied adhesive bonding in science and technology. IntechOpen, 2018. DOI: 10.5772/intechopen.68926

[12] Aubert JH. Thermally removable epoxy adhesives incorporating thermally reversible Diels-Alder adducts. Journal of Adhesion, 2003;79:609-616. DOI: 10.1080/00218460309540

[13] Das S, Samitsu S, Nakamura Y, Yamauchi Y, Payra D, Kato K, Naito M. Thermo-resettable cross-linked polymers for reusable/removable adhesives. Polymer Chemistry. 2018;9:5559-5565. DOI: 10.1039/C8PY01495G

[14] Turkenburg DH, van Bracht H, Funke B, Schmider M, Janke D, Fischer HR. Polyurethane adhesives containing Diels-Alder based thermoreversible bonds. Journal of Applied Polymer Science. 2017;134:1-11. DOI: 10.1002/app.44972

[15] Sounders JH, Frisch KC. Polyurethanes, Chemistry and Technology, Part I: Chemistry. Interscience, 1962. DOI: 10.1002/app.1964.070080342

[16] Lepene B.S, Long TE, Meyer A, Kranbuehl DE. Moisture-curing kinetics of isocyanate prepolymer adhesives. Journal of Adhesion. 2002;78:297-312. DOI: 10.1080/00218460210934

[17] Sridhar LM, Oster MO, Herr DE, Gregg JBD, Wilson, JA, Slark AT. Re-usable thermally reversible crosslinked adhesives from robust polyester and poly(ester-urethane) Diels-Alder networks. Green Chemistry. 2020;22:8669-8679. DOI: 10.1039/d0gc02938f

[18] US patent 8,734,939, 2014. Henkel Corporation

[19] Brancart J, Verhelle R, Mangialetto J, Van Assche G. Coupling the microscopic healing behaviour of coatings to the thermoreversible Diels-Alder network formation. Coatings. 2019;9:13. DOI: 10.3390/coatings9010013

[20] Li L, Chen X, Jin K, and Torkelson JM. Vitrimers designed both to strongly suppress creep and to recover original cross-link density after reprocessing: Quantitative theory and experiments. Macromolecules. 2018;51:5537-5546. DOI: 10.1021/acs.macromol.8b00922

[21] Czech Z, Wojciechowicz M. The crosslinking reaction of acrylic PSA using chelate metal acetylacetonates: European Polymer Journal. 2006;42: 2153-2160. DOI: 10.1016/j.eurpolymj.2006.03.022

Chapter 5

Pyrolysis of Furfural Residues and Possible Utilization Pathway

George Ngusale

Abstract

The manuscript attempts to understand the evolution of NOx precursors: NH_3 and HCN from Pyrolysis of furfural residue (FR). The pyrolysis process was carried out in a thermogravimetric analyzer (TGA) coupled to Fourier-transform infrared (FTIR) spectrometer. The combination revealed insightful information on the evolution of NH_3 and HCN. This could help us better understand the characteristics of FR derived from furfural production especially with regard to NH_3 and HCN. Nitrogen is considered a minor component in biomass wastes; in this study nitrogen content is about 0.57%. However, the pollution potential poised by low nitrogen content is huge through both direct and indirect processes. Thus, this study presents results that were found with regard to FR pyrolysis in pure nitrogen environment. At the heating rate of 40°C/min^{-1}, the only NOx precursor detected was HCN at 713 cm^{-1} as per the database provided by National Institute of Standards and Technology (NIST). NH_3 was not detected. The particle size of FR used ranged between 0.15–0.25 mm.

Keywords: furfural residues, pyrolysis, NOx precursors, TGA, FTIR

1. Introduction

In China, there exists abundant biomass in terms of agricultural crop residues that until now remain unutilized or under-utilized as shown **Figure 1**.[1] The crop residues range from rice husks/straws, wheat husks, corn cobs and many others. However, crop residues such as corn cob are greatly used in furfural industries for furfural production. Furfural production produces waste, herein referred to as Furfural Residue.

Furfural residues are in abundant supply in China [1]. It is estimated that a furfural processing plant with an annual furfural production of 1000 t produces about 13000 t/a of FR [2]. The residuals are derived from corncob through an acid hydrolysis process as shown in **Figure 2**. The process eliminates hemicellulose component retaining the cellulose and lignin components, with probably a few traces of hemicellulose [3].

With the increasing demand for energy in China, FR has been undergoing research to determine its utilization in various aspects such as ethanol production, soil conditioning to mention but a few [2, 3]. The research has been motivated by the great need to avoid open burning of FR in open fields. Worldwide, the open burning and/or disposal of any given waste or residue has been known to be a major

[1] http://dimsums.blogspot.com/2010/12/69-of-crop-residues-utilized.html. 2010. [accessed 2016-02-16].

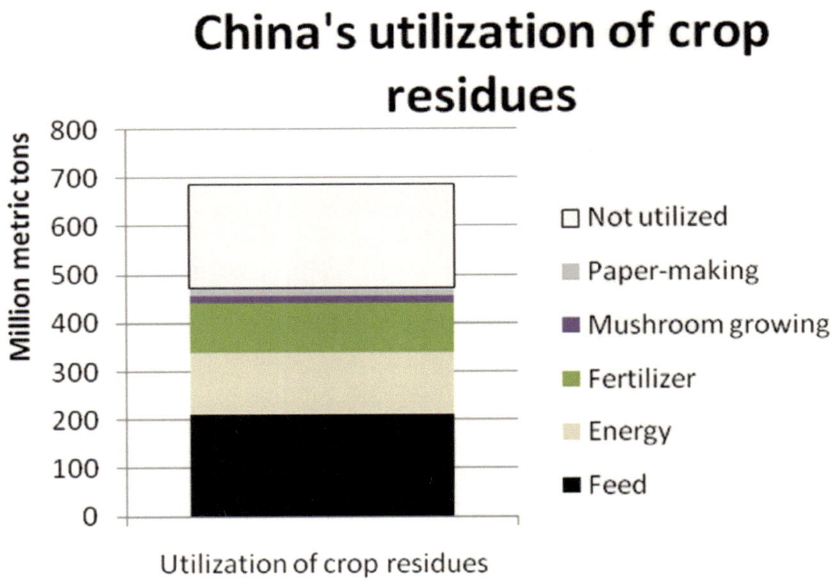

Figure 1.
Utilization of crop residues in China.

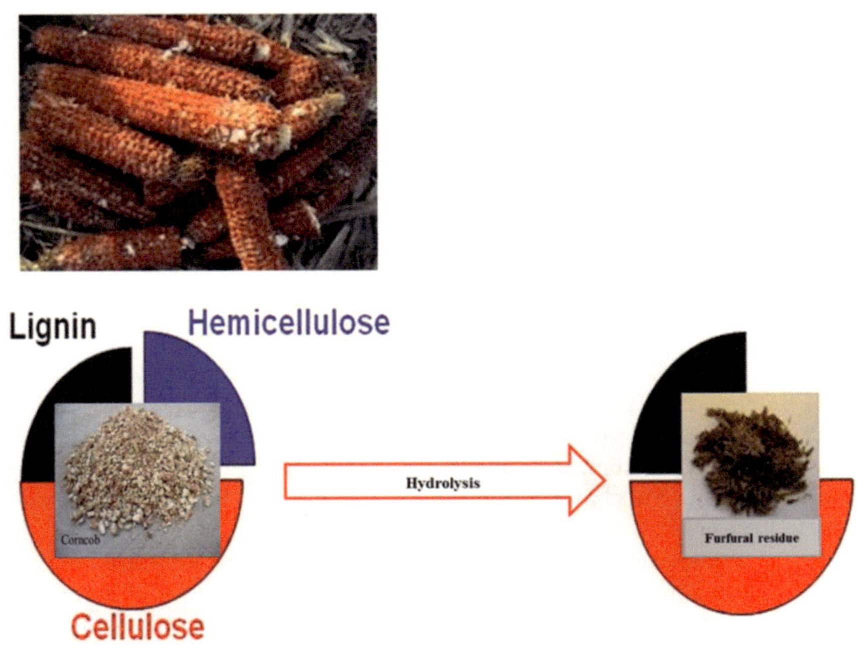

Figure 2.
Furfural residue formation.

environmental pollutant [4, 5]. On the other hand, limited literature currently exists on thermal-chemical utilization of FR for energy production.

However, for FR to undergo any thermo-chemical process (pyrolysis, gasification or combustion), knowledge on fuel-N is crucial. Fuel-N is normally converted into NOx which is environmentally harmful [6, 7]. In this regard, the manuscript considers the formation of NOx precursors mainly NH_3 and HCN. Previous studies have indicated that different types of biomass have different intrinsic properties especially nitrogen functionalities [8]. Furthermore, some studies have found that biomass pyrolysis yields more NH_3 more than HCN while others the vice-versa is true [7]. Therefore, this study aims to provide information on conversion of FR-Nitrogen during pyrolysis. This would offer a platform for comparability with previous studies on pyrolysis of other biomass wastes [9]. The results so obtained would go a long way in investigating the mechanism of NOx precursors for specific biomass wastes during pyrolysis process. The study used TG coupled to FTIR to investigate NH_3 and HCN. Given the low nitrogen content in FR (about 0.57%), the only NOx precursor detected was HCN at 713 cm^{-1} as per the database provided by NIST[2] [10].

2. Experiment

The proximate and ultimate analyses on FR carried out at Shanghai Jiao tong University are as shown in **Table 1**. Thereafter, about 5 mg of the samples whose diameter range from 0.15–0.25 mm were measured and then loaded in an inert platinum crucible placed on a balance. According to one of the manufacturers' Engineer of the TGA Q5000 (in Shanghai, China), no more than 5 ± 0.5 mg should be loaded to ensure decomposition takes place in the kinetic regime. Then, the correct positioning of the sample holder was ensured to avoid or minimize significant

Proximate analysis (wt. %)				Ultimate analysis (wt.%, dry basis)				
Fixed carbon[a]	**Volatile**	**Ash**	**Moisture**	s	H	O	N	S
39.13	23.31	6.95	30.61	46.51	4.94	35.09	0.51	2.09

[a]*By difference.*

Table 1.
Proximate and Ultimate analysis of FR.

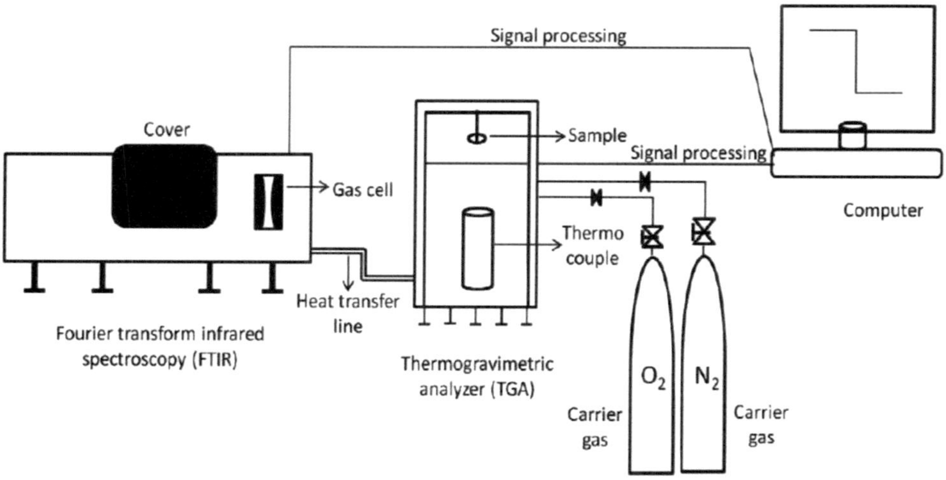

Figure 3.
Schematic diagram of the TGA coupled to FTIR spectrometer [13].

[2] National Institute of Standards and Technology (NIST) chemistry WebBook Database of organic chemistry compounds.

temperature shifts/errors that would have occurred [11]. Pyrolysis was carried out at a heating rate of 40°C/min from ambient room temperature to final temperature of 1000°C. Though the low heating rate selected does not correspond to realistic thermal conditions, the value is necessary for analysis and understanding the experimental results. Nitrogen gas was used as the purge gas to sweep the pyrolysis gases and prevent secondary reactions and tar deposition on the FR sample while providing an inert atmosphere. The flowrate was set at 50 ml/min as the value does not have significant effect on amount of collected gaseous products [5]. Evolved gases from TGA were transmitted to a Bruker Tensor FTIR (vertex 70) spectrometer via a heated transfer line kept at a constant temperature of 180–190°C to prevent the condensation of less volatile compounds [12]. FTIR spectra were collected with 1 cm^{-1} resolution, in the range of 4000-600 cm^{-1} IR absorption band. A schematic setup of the complete TGA coupled to FTIR spectrometer is shown in **Figure 3**.

3. Results and discussion

3.1 Fuel properties

From **Table 1**, the percentage results of both proximate and ultimate analysis of FR were used to calculate the empirical formula (EF): $CH_{1.2746}O_{0.5659}N_{0.009}S_{0.0169}$. Then, using the EF, the value of the stoichiometric weight air-to-fuel ratio, Rs was obtained as 5.62. The formula for Rs assumes complete conversion to water vapors and carbon dioxide. Rs value for FR is approximately the same as that obtained for nutshells (Rs = 5.6) [14]. Similarly, the moisture content is high at 30.61% while nitrogen content is at 0.51%. This nitrogen content is within the range measured by other researchers [2, 15].

3.2 TGA—FTIR

TGA coupled to FTIR spectrometer has been known to provide insightful information on the composition of gaseous products evolved from the pyrolysis of solid fuels [13, 16]. In this study, a heating rate of 40°C/min was selected high enough to ensure the quality of the FTIR data obtained [12].

3.2.1 TGA curves for FR

Figure 4 shows TGA curves for both weight loss and the corresponding derived weight of FR. Owing to limited literature on FR pyrolysis, the Authors attempted to learn some relevant information on FR [3, 17]. The information helped deduce the thermal decomposition behavior of FR. The initial temperature rise necessitated moisture release from FR. The moisture released contributes about 10% of weight loss. Volatile matter started to be released from FR at temperatures of about 239°C. When the temperature reached about 748°C, more than 65% of the weight of FR was released under pure pyrolysis condition (using nitrogen as the medium). A final high temperature of 1000°C was selected.to ensure complete FR degradation. This might be due to higher thermo-stability of lignin in FR due to occurrence of more condensed polymeric structures in its composition [3]. Similarly, derived weight curve shows four peaks: one major peak and three minor ones. First minor peak occurs between 62 and 93°C with a maximum decomposition rate of 0.88%/°C. This indicates the released moisture from within

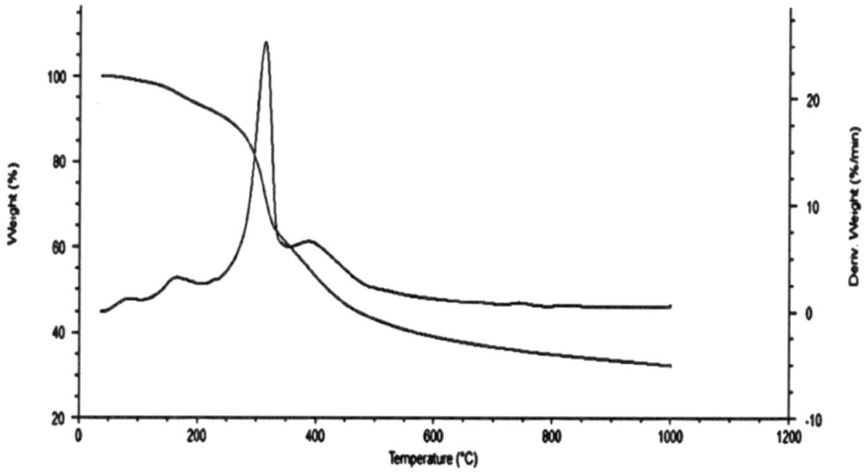

Figure 4.
TGA curves for both the weight loss and derived weight of FR.

FR material. The second peak occurs between 137 and 195°C with a maximum decomposition rate of 3.1%/°C, indicating the degradation of hemicellulose traces left in FR material after hydrolysis. The third peak is the major one as it represents the greatest weight loss at temperatures between 275 and 334°C with a maximum decomposition rate of 25.2%/°C. Last but not least, fourth peak is attributed to lignin degradation in FR material at a broad temperature range of 369–416°C with a maximum decomposition rate of 6.53%/°C. This could be as a result of hemicellulose extraction that tampered with the lignin content or rather a hypothesis that lignin undergoes a phase transition [18]. Some studies have found that at high temperatures (>400°C) the aromatic rings of char matrix in lignin are rearranged [19, 20]. However, the occurrence of the different peaks solely depends on the percentages of hemicellulose, cellulose and lignin contained in the original material, which in this case is FR [21].

3.2.2 FTIR analysis of HCN and NH_3 in the pyrolysis gas composition

The gaseous emissions measured with the FTIR system are as shown in **Figures 5** and **6**. **Figure 5** shows the three-dimensional (3D) diagram corresponding to FR pyrolysis while **Figure 6** shows IR spectrum of pyrolysis products obtained at the maximum evolution rate for each decomposition FR.

The system has built-in calibrations for various gaseous emissions. Since the focus of this study was on HCN and NH_3, the other emissions were not identified and analyzed. The wave number ranges for HCN and NH_3 selected were 712–714 cm^{-1} and 930–966 cm^{-1} respectively [9, 22]. At these wave numbers no major moisture content interference especially for HCN apart from the normal bending vibrations in the fingerprint region.

HCN, NH_3 and other nitrogenous species are known to be present in low amounts [23–25]. **Figure** 7 shows HCN released from pyrolysis of FR in nitrogen atmosphere. The initial HCN released began at a temperature of about 152°C and then increased sharply to a maximum temperature of 353°C. However, a sharp rise in HCN released was evident within the temperature range of 286–386°C then slightly stabilized up to 420°C before gradual decline up to 822°C. Thereafter,

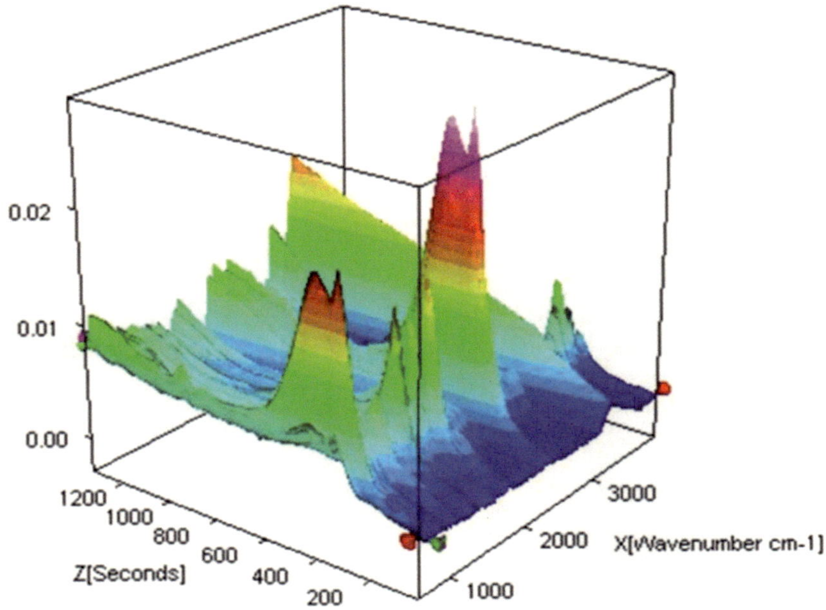

Figure 5.
3D IR spectrum of Furfural residues.

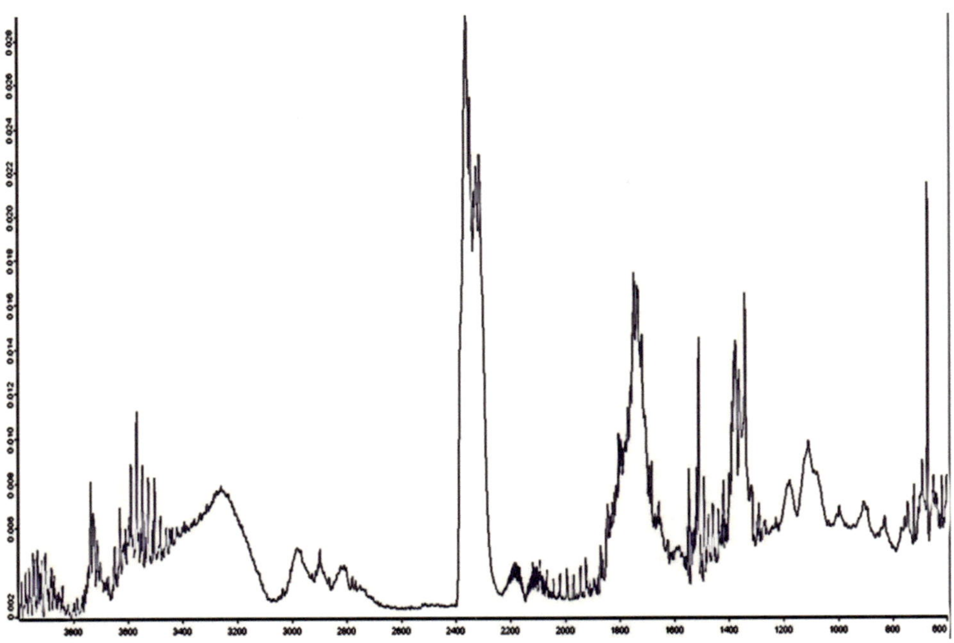

Figure 6.
IR spectrum of pyrolysis products obtained at the maximum evolution rate for each decomposition Furfural residues.

a gradual rise took place up to 1000°C. NH_3 released could not be conclusively detected as per the NIST webBook. The study however, showed similar trends of previous detection of HCN, NH_3 and other nitrogenous species with regard to pyrolysis of other biomass wastes [26, 27].

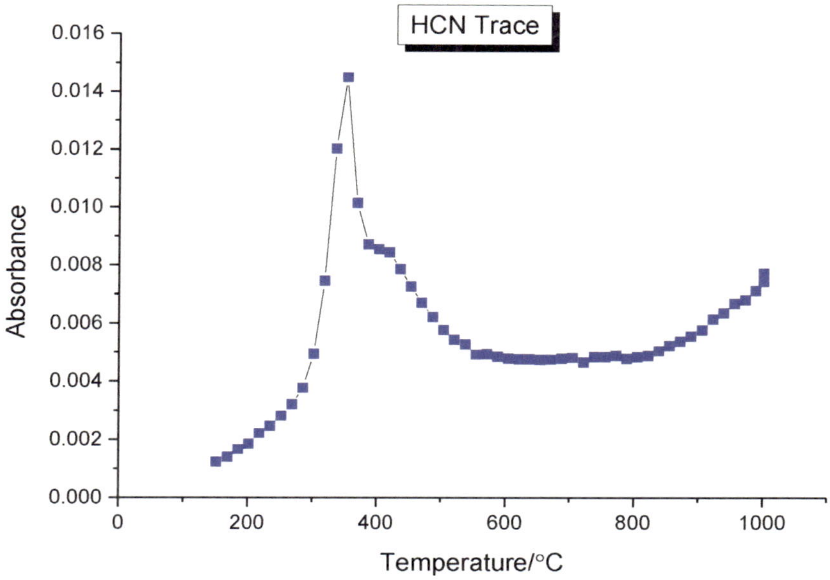

Figure 7.
Release of HCN in nitrogen atmosphere.

4. Conclusion

The evolution characteristics of NH_3 and HCN released from the pyrolysis of FR were investigated by TGA-FTIR. The pyrolysis temperature range of 200–485°C led to a maximum release of HCN at 353°C. At the heating rate of 40°C/min^{-1}, the only NOx precursor detected was HCN at 713 cm^{-1} as per the database provided by NIST. NH_3 was not detected while HCN was released from pyrolysis of FR in nitrogen atmosphere. The particle size of FR used ranged between 0.15–0.25 mm.

Also, percentage results of both proximate and ultimate analysis of FR obtained an empirical formula (EF): $CH_{1.2746}O_{0.5659}N_{0.009}S_{0.0169}$. Then, using the EF, the value of the stoichiometric weight air-to-fuel ratio, Rs, was 5.62.

As the focus of this study could not factor the effect of both varying particle size and heating rate, further research is recommended.

Author details

George Ngusale
Jaramogi Oginga Odinga University of Science and Technology, Bondo, Kenya

*Address all correspondence to: kavulavu06@yahoo.co.uk

References

[1] Sun Y, et al. Study on conversion process for furfural residue manufacture to ethanol by simultaneous saccharification and fermentation. Mod Chem Ind, 2008. 28(12): p. 48-52.

[2] Liu B, Girisuta B, Heeres H. Research progress on furfural residues recycling: A literature review. in Environmental Engineering and Applications (ICEEA), 2010 International Conference on. 2010. IEEE.

[3] Bu L, et al. Comparative characterization of milled wood lignin from furfural residues and corncob. Chemical Engineering Journal, 2011. 175: p. 176-184.

[4] Malkow T. Novel and innovative pyrolysis and gasification technologies for energy efficient and environmentally sound MSW disposal. Waste management, 2004. 24(1): p. 53-79.

[5] Dalai A.K, et al. Gasification of refuse derived fuel in a fixed bed reactor for syngas production. Waste management, 2009. 29(1): p. 252-258.

[6] Ogawa M, Yoshida N. Nitrous oxide emission from the burning of agricultural residue. Atmospheric Environment, 2005. 39(19): p. 3421-3429.

[7] Ren Q, et al. Formation of NO_X precursors during wheat straw pyrolysis and gasification with O_2 and CO_2. Fuel, 2010. 89(5): p. 1064-1069.

[8] Ren Q, et al. Effect of mineral matter on the formation of NO_X precursors during biomass pyrolysis. Journal of analytical and applied pyrolysis, 2009. 85(1): p. 447-453.

[9] Di Nola G. Biomass fuel characterization for NOx emissions in cofiring applications. 2007.

[10] NIST. http://webbook.nist.gov/cgi/cbook.cgi?Scan=cob10480&Type=IR. HCN-IR spectrum. [Accessed:2016-05-03].

[11] Stenseng M, Jensen A, Dam-Johansen K. Investigation of biomass pyrolysis by thermogravimetric analysis and differential scanning calorimetry. Journal of analytical and applied pyrolysis, 2001. 58: p. 765-780.

[12] Gómez-Siurana A, et al. TGA/FTIR study of tobacco and glycerol–tobacco mixtures. Thermochimica Acta, 2013. 573: p. 146-157.

[13] Parshetti G.K, et al. TGA–FTIR investigation of co-combustion characteristics of blends of hydrothermally carbonized oil palm biomass (EFB) and coal. Fuel Processing Technology, 2014. 118: p. 228-234.

[14] Di Blasi C, Signorelli G, Portoricco G. Countercurrent fixed-bed gasification of biomass at laboratory scale. Industrial & engineering chemistry research, 1999. 38(7): p. 2571-2581.

[15] Wang Qing, Hou Fengyun, Sun Donghong, Su Guiqiu, Sun Jian. Research on the pyrolysis characteristics of furfural residue (Doctoral dissertation), 2004.

[16] Yuzbasi N.S, Selçuk N. Air and oxy-fuel combustion characteristics of biomass/lignite blends in TGA-FTIR. Fuel Processing Technology, 2011. 92(5): p. 1101-1108.

[17] Sun S.-N, et al. Alkaline and Organosolv Lignins from Furfural Residue: Structural Features and Antioxidant Activity. BioResources, 2013. 9(1): p. 772-785.

[18] Gunarathne D.S, et al. Gasification Characteristics of Hydrothermal Carbonized Biomass in an Updraft Pilot-Scale Gasifier. Energy & Fuels, 2014. 28(3): p. 1992-2002.

[19] Antal M.J.J, Varhegyi G. Cellulose pyrolysis kinetics: the current state of knowledge. Industrial & engineering chemistry research, 1995. 34(3): p. 703-717.

[20] Grønli M, Antal M.J, Várhegyi G. A round-robin study of cellulose pyrolysis kinetics by thermogravimetry. Industrial & engineering chemistry research, 1999. 38(6): p. 2238-2244.

[21] Chen W, et al. Updraft Gasification of Mesquite Fuel Using Air/Steam and CO2/O2 Mixtures. Energy & Fuels, 2013. 27(12): p. 7460-7469.

[22] Kruse J, et al. TG–FTIR, LC/MS, XANES and Py-FIMS to disclose the thermal decomposition pathways and aromatic N formation during dipeptide pyrolysis in a soil matrix. Journal of analytical and applied pyrolysis, 2011. 90(2): p. 164-173.

[23] Tao L, et al. TG–FTIR characterization of pyrolysis of waste mixtures of paint and tar slag. Journal of hazardous materials, 2010. 175(1): p. 754-761.

[24] Fu P, et al. FTIR study of pyrolysis products evolving from typical agricultural residues. Journal of analytical and applied pyrolysis, 2010. 88(2): p. 117-123.

[25] Ferrasse J, et al. Chemometrics as a tool for the analysis of evolved gas during the thermal treatment of sewage sludge using coupled TG–FTIR. Thermochimica Acta, 2003. 404(1): p. 97-108.

[26] Giuntoli J, et al. Quantitative and kinetic TG-FTIR study of biomass residue pyrolysis: Dry distiller's grains with solubles (DDGS) and chicken manure. Journal of analytical and applied pyrolysis, 2009. 85(1): p. 301-312.

[27] Hardy A, et al. Study of the decomposition of an aqueous metal–chelate gel precursor for (Bi, La)< sub> 4</sub> Ti< sub> 3</sub> O< sub> 12</sub> by means of TGA–FTIR, TGA–MS and HT-DRIFT. Thermochimica Acta, 2003. 397(1): p. 143-153.

Chapter 6

Furanic Rigid Foams, Furanic-Based Bioplastics and Furanic-Derived Wood Adhesives and Bioadhesives

Antonio Pizzi and Anish Khan

Abstract

In this chapter, we discuss pure furanic foams and tannin-furanic foams as fire-resistant, environmentally friendly, rigid biofoams. We also examine furanic wood adhesives in which a major furan portion is coupled with either synthetics or bioadhesives. In the case of furanic wood bioadhesives, the formulations developed were 90–100% biosourced. Equally, furanic rigid plastics of considerable mechanical resistance have also been developed and applied to angle-grinder discs and automotive brakes with very encouraging results.

Keywords: furanic foams, tannin-furanic foams, tannin-furanic adhesives, tannin-furanic plastics, tannin-furanic films

1. Introduction

Biosourced furan derivatives such as furfural, furfuryl alcohol, and hydroxymethyl furfural have been a focus of research in the last 10 to 15 years in several different application fields. This chapter deals with three topics: (1) fire-resistant furan-based foams, (2) the co-reaction to prepare wood panel bioadhesives of furanic materials with renewable and environmentally friendly materials, and (3) the preparation of hard plastics by reacting different natural and environmentally friendly renewable materials with furanics. A considerable level of research activity has been recorded in all these areas.

2. Fireproof furanic rigid foams

2.1 Tannin-furanic foams by chemical exothermal reactions

In this section we discuss the preparation of almost totally biobased tannin-furanic foams via expansion/blowing of the foam by chemical exothermal reactions caused by the heat generated under acid conditions of the self-condensation of furfuryl alcohol. Then, we discuss tannin-furanic foams in which isocyanate is added in the minority to the tannin-furanic mix. We also address the applicability of these foams to isocyanate-based polyurethane foam factories. Synthetic isocyanate-based polyurethane foams, even those using biopolyols, are not generally fire resistant unless some fire-retardant is used. Although the foams presented here are intrinsically fire resistant, like phenolic foams, but without their pollution characteristics.

Pure furanic foams are prepared by polycondensation of furfuryl alcohol under acid conditions [1–3]. Furanic foams are commonly used in foundries, because of their high resistance to heat and their relatively low cost, to bind the sand of molds or cores for casting engine heads or other kinds of steel tools [4, 5]. A study on the formation of pure furanic foams and the conservation and modification of their structure after carbonization is described in [6]. Ambient temperature catalysis of furfuryl alcohol with para-toluen sulphonic acid (pTSA) is the method used to prepare pure furanic foams. This research shows that the furfuryl group is the main repeating unit/motive from which derive the whole variety of structures observed in the polymer network formed (**Figures 1** and **2**), with structures shown in **Figure 2** been present.

The same work [6] studied the type of structures that remain in a furanic foam after carbonization at 900°C. The research shows that many polynuclear aromatic hydrocarbons are present after carbonization (**Figure 3**). The average molecular weight of the fragments increases during carbonization because of the rearrangement of the furanic structures. Gasification during carbonization makes the signal

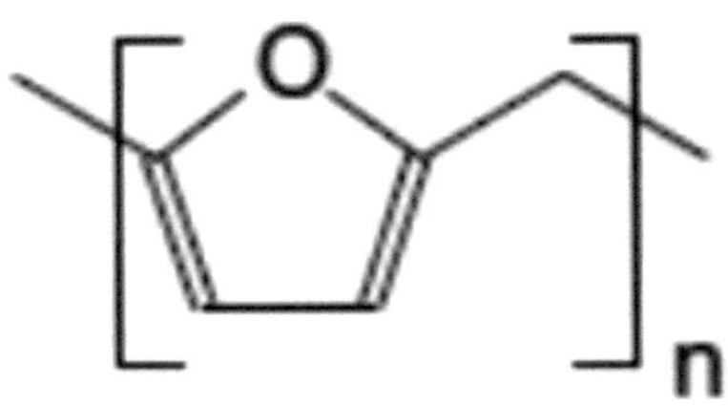

Figure 1.
Structure of the linear furanic oligomers formed by the self-condensation of furfuryl alcohol.

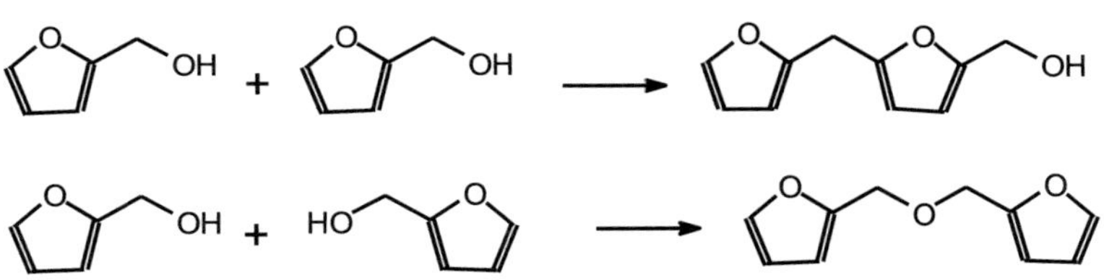

Figure 2.
Methylene and methylene ether bridges linking furanic nuclei in furanic oligomer structures found in linear oligomers from the self-condensation of furfuryl alcohol (FA). Top: reaction of the $-CH_2OH$ *of FA with the furanic ring of a second FA molecule. Bottom: reaction of the* $-CH_2OH$ *of FA with the* $-CH_2OH$ *of another molecule of FA. Both reactions are obtained by elimination of water.*

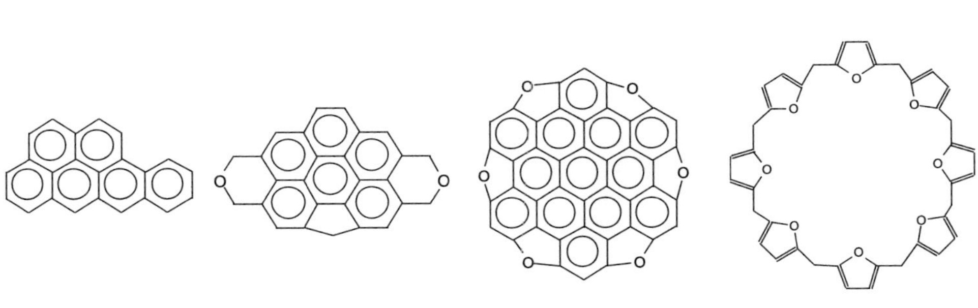

Figure 3.
Example of rearranged structures formed during carbonization of furanic foams [6].

of the pTSA catalyst derivates disappear by degrading to toluene and SO2 not surviving carbonization. However, certain furanic oligomers survive carbonization; these are mostly cyclic compounds with 4–6 or more furan rings. Thus, even if most of the constituents are transformed to more stable aromatic structures, some of the starting chemical species survive intact or partly transformed to carbonization even if most structures are converted to more stable aromatic structures (**Figure 3**). This shows the stability of some furanic oligomers that are not degraded or rearranged by carbonization. Molecular mechanics calculation of their relative energies appeared to confirm that these structures are cyclic furanic oligomers.

Tannin-furanic foams were mentioned for the first time in the literature in the early 1970s when Grey, Roux, Pizzi, and Ryder developed a foam formulation in South Africa [7]. This formulation had some severe problems and its performance was unacceptable. Moreover, industry did not appear to have any interest in the foam, as there was more focus on the dominant synthetic oil-derived foams at the time. In addition, the public opinion of the relative cost structure/performance relationship of these foams and of the biomaterials used also appeared to be unfavorable. The first tannin-furanic biofoam formulation that appeared to work well was published in the literature in 1994 by Meikleham and Pizzi [8]. Nonetheless, even then there was no interest in these materials; interest in them materialized only in the late 2000s [9].

Ambient temperature, self-blowing tannin-furanic foams were the first researched for a relatively long period, these being chemically foamed and set by the exothermic acid self-condensation of furfuryl alcohol (**Figure 4**). In earlier times, diethyl ether was used as a blowing agent [8]. The foams prepared with this early research were either catalyzed by an acid or a base but showed characteristics and performance comparable to synthetic phenolic foams. The liquid polymer phase was a tannin-formaldehyde resin. Foaming occurred by the forced evaporation of a physical blowing agent, while cross-linking rendered the foams dimensionally stable and with the goal target density. Acid-catalyzed foams expanded by evaporating the blowing agent due to the heat-surge agent produced by the self-condensation of furfuryl alcohol. Tannin-furan copolymers were so obtained. No toxic gasses on these foams' carbonization were detected [9, 10]. This formulation worked, but when this research was again started up in the late 2000s several problems remained that needed to be solved. These were (1) the elimination of formaldehyde, used up to then to improve cross-linking and (2) the elimination of diethyl ether by substituting it with a less volatile and less dangerous blowing agent. In-depth characterization of these foams ensued. Different condensed tannins, namely, mimosa tannin bark extract, pine bark tannin, and quebracho tannin wood extract, were coupled to furfuryl alcohol as foam building blocks. Hydroxymethylated lignin up to a level of 20% by weight, and even smaller proportions of polyurethane, isocyanate [11], and industrial surfactant [9], were added

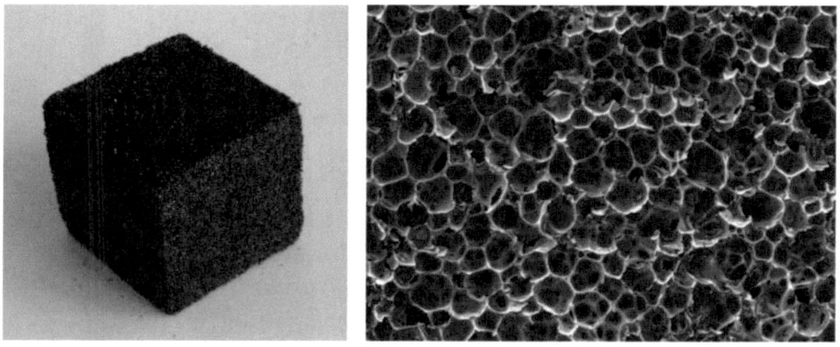

Figure 4.
An example of a tannin-furanic rigid foam (left) and of its structure as observed at the scanning electron microscope (right).

successfully to modify these foams. Physical tests such as water absorption, compression resistance, direct flame behavior, and measure of foam cells' dimensions were carried out [9]. These foams were chemically characterized by 13C-NMR analysis.

Equally, these tannin-furanic foams' resistance to fire and chemicals (e.g., resistance up to 1200°C and higher), absorption of and resistance to various liquids (e.g., ethanol, 33% sulphuric acid, and organic and mineral acids), permeability, thermal conductivity (e.g., between 0.024 and 0.044 W/m K), and mechanical (compressive and tensile) strength were tested. Boric acid and/or phosphoric acid were added to modify the foams' structure while improving substantially their fire resistance (**Figure 3**). Rigid foams resisting well to strong acid, bases, and solvents were prepared with these early formulations [12]. High affinity for water, but limited affinity for organic solvents, was also put into evidence. Slightly anisotropic mechanical properties were measured. These foams were brittle in tension and compression, but their thermal conductivity and mechanical performance fully compared with those of synthetic phenolic foams. X-ray microtomography was also used to examine these early foams [13–16]. This provided additional and useful information regarding their physical characteristics such as porosity, pore size distribution, proportion of open and closed cells, connectivity, and tortuosity.

These foams garnered considerable interest for their phenomenal fire resistance and hence their heat insulation potential [16]. Even when exposed to a flame 1200°C or higher, they do not burn for whatever length of time. The red color induced on the area where the flame is applied is automatically and immediately self-extinguishing on subtracting the flame. These tannin-furanic foams only begin to decompose at 3000°C. We will discuss the multiple uses for which they were tested later in the chapter. These include, for example, thermal insulating materials, fire-resistant filling for hollow doors and wood sandwich structures [17], green acoustic absorption materials [18], support for cut flowers [19] and hydroponic cultures [19], and flexible and rigid polyurethane-tannin-furanic mixed foams [20, 21].

The substitution of diethyl ether as a blowing agent with a much safer solvent was the first modification introduced in these tannin-furanic foams. The blowing agent chosen was n-pentane in synthetic phenolic foams. A necessary formulation rebalancing ensued, as pentane boils at a temperature higher than diethyl ether.

The first necessity was to eliminate formaldehyde from the formulation, formaldehyde having been declared unsuitable for sanitary reasons [22, 23]. With the elimination of formaldehyde, the new foams that were obtained presented lower density, thermal conductivity, hydrophilicity, and brittleness, and thus greater flexibility than the first generation of tannin-furanic foams. Formaldehyde was replaced with more furfuryl alcohol and a greater proportion of blowing agent, significantly improving the characteristics previously described [24, 25]. **Table 1** shows the characteristics of these foams regarding their range of compressive strength and thermal conductivity as a function of their apparent density. **Table 1** shows that as the apparent foam density increases the compressive strength sharply improves; however, the thermal conductivity increases, which is less acceptable as regards insulation.

Next, both formaldehyde and solvent (pentane) were eliminated, rendering the foams 98% "green." Comparison of kinetic curves describing the simultaneously measured foams' expansion, hardening, temperature, and pressure variation as a function of time illustrated the differences in process and foaming parameters as a function of time by the differences in formulation between the experimental and control foams and optimization of the foaming and hardening parameters involved [26, 27].

A first encouraging attempt to prepare elastic tannin-furanic foams occurred at this time [28]. In this first successful approach, flexible tannin-furanic foams, rather than the rigid ones prepared up to then, were obtained by the addition of

Apparent density (g/cm^3)	Compressive strength (MPa)	Thermal conductivity (W m^{-1} K^{-1})
0.016	0.03	0.024–0.030
0.040–0.080	0.12–0.45	0.040–0.050
0.10–1.30	0.65–1.10	0.050–0.060

Table 1.
Typical range of physico-mechanical values for tannin-furanic-based foams [12, 14, 30, 33, 54].

glycerol as an external (unreacted) plasticizer. The choice of glycerol was dictated by its high boiling temperature, lack of evaporation, and lack of toxicity. Flexibility and spring-back of these experimental foams when subjected to a cyclic compression force followed by spring-back and compression again was quantified by both thermomechanical analysis at different temperatures as well as by compression/spring-back hysteresis cycle tests in a universal testing machine. Tannin-furanic foams with formaldehyde and no glycerol reached a stress plateau indicative of structure crushing. Tannin-furanic foams without both formaldehyde and glycerol become very fragile, brittle, and rigid just two months after their preparation. They also show structure crushing with ageing. Tannin-furanic foams with no formaldehyde but with glycerol remain instead equally and truly flexible in time [27, 28].

Furthermore, open cell foams obtained by the simultaneous co-reaction of condensed flavonoid tannins with an alkoxylated fatty amine and polymeric diphenylmethane isocyanate yielded highly flexible/elastic polyurethane foams [28]. Copolymerized amine/isocyanate/tannin oligomers were identified by ^{13}C NMR and MALDI-TOF spectroscopy. In general, between 30% and 50% of natural tannins is added to the components used to polymerize the polyurethane. The characteristic of these new, partially biosourced polyurethanes is that the presence of the tannin slows down burning; some of them can be made flame self-extinguishing and if burning they neither flow nor sprinkle flaming material around, contrary to what occurs with normal polyurethanes. This limits transmitting the fire to other materials in the same environment. Cyclic compression tests were carried out showing that after 50 cycles foam recovery was more than 80%.

Hyperbranched poly (acylamide-ester) polyol synthesized reacting in one step succinic anhydride with diethanol amine was also used to modify tannin-furanic foams [29]. Glutaraldehyde was reacted with the hyperbranched poly (acylamide-ester) polyol to acetalize it, and the dendrimer so prepared was used to modify the tannin-based foams. It was found that the compression strength of the tannin-furanic foam improved by 36.6% with the addition of 3.5 wt% of acetalized poly (acylamide-ester) polyol without affecting the other foam properties.

Pine bark tannins are much more reactive than mimosa and quebracho tannins experimented with up to 2012. Pine bark tannin-furanic foams were prepared for the first time in 2013 [30–33]. The tannin-furanic foam formulations underwent fundamental changes due to the greater pine tannin reactivity. This had to be implemented to coordinate foam hardening, reaction exotherm, and solvent blowing to obtain a rigid foam. This work was achieved using the FOAMAT, an equipment able to simultaneously monitor during foaming the variation of temperature, pressure, velocity, and dielectric polarization. This allowed for determining the function of the surfactant (castor oil ethoxylate) and the plasticizer (polyethylene glycol) during foam formation and thus to monitor their polymerization, expansion, hardening, and shrinkage. Foam density and its physical properties were found to be either surfactant- or plasticizer-controlled in this research work. Foams presenting a homogeneous microstructure were obtained with castor oil ethoxylate and polyethylene glycol. However, polyethylene glycol made the foams more elastic but with lower shrinkage. Pine tannin-furanic foams both with and without formaldehyde were also prepared and tested to determine their stress–strain curves, thermal conductivity, Young's modulus, compression strength, densification, densification rate, and energy absorbed under compression. These pine tannin-furanic foams with formaldehyde had properties similar to mimosa tannin-furanic foams. At very low densities, mimosa foams are more mechanically resistant than pine foams. Mimosa foams with formaldehyde have a greater Young's modulus less than 0.10 g cm^{-3} and a greater compressive strength less than 0.14 g cm^{-3} than pine tannin foams. Pine tannin-furanic foams without formaldehyde were more elastic and had lesser mechanical strength. However, on a comparative test, pine tannin foams are a better insulation material with an average thermal conductivity of 0.030 W/m/K for pine foam without formaldehyde, 0.034 W/m/K for pine foam with formaldehyde, and 0.037 W/m/K for mimosa tannin foam with formaldehyde at a density of 0.031 g cm^{-3}.

The reformulation undertaken for these pine tannin-furanic foams allowed to develop such foams for the whole class of very reactive procyanidin tannins and not only different species of pine tannins [30–33] such as spruce tannins [34–37], and others. Pine tannin-furanic foams free of any aldehyde, and of formaldehyde, have also been developed, but their main drawback is their lower resistance to compression. Consequently, formaldehyde-free pine tannin-furanic rigid foams were successfully obtained by using non-volatile aldehydes [22, 23], namely glyoxal or glutaraldehyde, as alternative non-toxic hardeners [30–32]. All the open-cell pine tannin-furanic foams and mimosa–/quebracho-type tannin-furanic foams have also yielded medium and high frequencies (1000–4000 Hz) and good sound absorption/acoustic insulation with acoustic absorption coefficients of 0.85–0.97 [17]. They were better than polyurethane foams, melamine foams, fiberglass, and mineral wool acoustic insulations within this frequency range [17]. Their acoustic absorption coefficient decreased to 0.40–0.60 at lower frequencies of 250–500 Hz.

Tannin-furanic foams have shown typical characteristics comparable to synthetic commercial foams as light porous materials. Open-cell foams result in better sound absorption with thicker samples performing better in the medium frequency range.

Surface friability of tannin-furanic foams was a drawback for such potential applications, but this problem was also solved. A second main drawback is absorption of water within the foam itself. Both these drawbacks have been eliminated or at least minimized by adding to the formulation a small percentage of an oil-grafted tannin. The fatty chains introduced in the foam markedly decreased foam friability and increased water repellency in the foam's body [38]. Also, adding small amounts of soy protein hydrolysate decreased surface friability of these foams [18].

Lightweight sandwich panels with a tannin-furanic foam core and wood veneers or hardboard thin panels as surfaces bonded on to the foam core were also prepared (**Figure** 5) [39, 40].

As procyanidin tannins are the world's predominant potential source of condensed tannins the development outlined for pine tannins-furanic foams are of considerable importance as they allow any future diffuse utilization of tannin foams anywhere.

The determinant parameters when designing new tannin-furanic foams have been clearly identified and codified [41]. Further progress in this field can be achieved by anyone who would care to follow these parameter guidelines.

Tannin-based carbon aerogel foams innovatively based on the ionic and radical autocondensation of tannins under alkaline conditions promoted by their reaction with silica and silicates [42–46] have also been prepared [47]. Upscaling to pilot plant level of the preparation of these types of foams has also been reported [48].

Recently, some more progress on the chemical analysis of this foam has also been made using Raman spectroscopy and attenuated total reflection–Fourier transform

Figure 5.
Foam cores sandwiched from top left to bottom right between surfaces formed of wood veneers, thick solid wood boards, thin plywood surfaces, thick and thin hardboard surfaces, and not sandwiched (just foam core).

infrared spectroscopy (ATR–FTIR) approaches [49, 50]. Research groups have also been active in the preparation processes of these tannin-furanic foams [51–53] as well on the range of different applications possible, the most notable being in the medical field [21]. For example, tannin-furanic foams can be used in medicine to form a tannin-hydoxyapatite scaffold of stem cells for bone reconstruction without using any synthetic materials [21].

3. Tannin foams by formation of mixed tannin-furanic and tannin-based polyurethanes

While phenolic foams can be clearly substituted to good effect with tannin-furanic foams, the market is particularly interested in the use of biobased polyurethane foams. This interesting situation came to the fore with an industrial plant trial for a plant where isocyanate had to be compulsorily used, otherwise the plant could not run. This was furthermore quite a sizeable polyurethane foam panels line (approx. 18 thousand tons/year). Mixed phenolic–polyurethane-type rigid foams were developed using tannin-furfuryl alcohol natural materials co-reacted with polymeric isocyanate in the proportions imposed by the limitations inherent to the continuous industrial plants for polyurethane foams and used in the plant trial [54]. Chemical analysis of the final foams identified several different copolymerization oligomers having been generated. Urethane linkages were generated by reaction of the isocyanate with two flavonoid tannin reactive sites, mainly at the flavonoid aliphatic hydroxygroup at C3, and less so on the phenolic hydroxygroups of the tannin flavonoid units. Urethane linkages also formed by isocyanate reaction with (1) glyoxal both alone or pre-reacted with the tannin, (2) the phenolsulfonic acid catalyst, and (3) with furfural. This latter reagent does react preferentially through formation of a methylene bridge with the A-ring of the flavonoid units of the tannin rather than reacting with the isocyanate to form urethanes [54]. All the materials appeared to have co-reacted to form a mix of urethane linkages and methylene bridges between all the main components used. Thus, the tannin, furfuryl alcohol, isocyanate, glyoxal, and even the phenolsulfonic acid catalyst formed a variety of mixed species linked by the two bridge types. Several mixed species constituted of 2, 3, and even 4 co-reacted different components were observed.

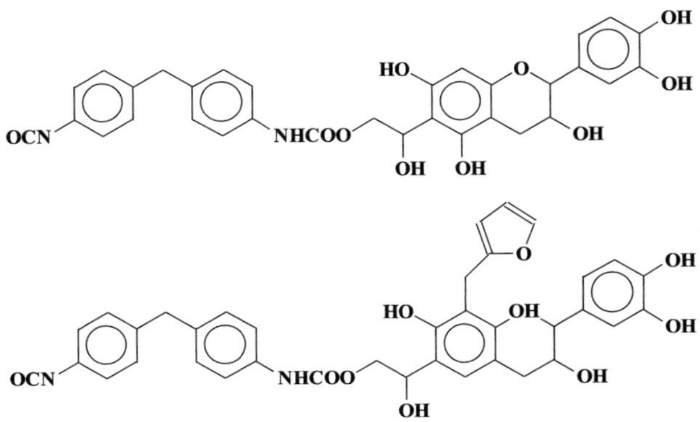

Figure 6.
Example of mixed tannin polyurethanes obtained by the reaction of the isocyanate group on the glyoxal groups pre-reacted with flavonoid tannin units. The reaction can be carried out simultaneously as well, as used under industrial conditions.

The more interesting result here, however, was that this approach was unusually different from the approach of oxypropylating tannins to render them more apt polyols for reaction with isocyanates [55, 56], hence using an additional reaction step. The unusual results [54] were especially interesting because they were obtained on an industrial plant line trial. Effectively, what occurred was that the glyoxal easily reacted with the tannin during the trial producing –OH groups much more easily approached by the isocyanate, thus forming a glyoxalated tannin polyurethane in a single step, which is a remarkably useful outcome [54]. Thus, species of the type shown in **Figure 6** were present.

The reaction of glyoxal with the tannin and then with isocyanates to form urethanes closely repeat the same reaction already used for wood adhesives but using the $-CH_2OH$ groups formed by the reaction of formaldehyde with tannins and with synthetic phenolic and amino resins [57–59]. As regards the fire resistance of these foams the preponderance of the tannin phenolic groups and furanic nuclei gives a certain level of fire resistance due to the inclusion of tannins in standard polyurethane formulations [20], but fire resistance is expected to be lower than the standard tannin-furanic foams described earlier in the chapter.

4. Wood and fiber adhesives

The potential of using tannin-furfuryl alcohol resin for biobased composites using vegetal fiber reinforcement has also been investigated [60–62]. Results showed that a mix of 54% furfuryl alcohol, 45% modified quebracho tannin extract, and 0.9% pTSA as a catalyst yields a resin with which one can prepare lightweight composites by working as bonding and solidifying matrix of a nonwoven flax fiber. The composite panels so prepared, once tested for tensile and flexural modulus and strength, water resistance, and thermo-degradation, presented good mechanical properties and a very short curing time in a hot press.

Tannin-furfuryl alcohol resins reacting under alkaline conditions to minimize self-condensation of furfuryl alcohol and force its reaction with tannins have proved to be another alternative for formaldehyde-free, environmentally friendly adhesives from renewable materials [63]. An indication of the reactivity of tannin with furfuryl alcohol to harden an adhesive composed of these two materials is given in **Figures 7** and **8** where it is shown that the mixture of the two materials gels at pH levels of less than 2–2.5 and greater than 8–9 according to the reactivity of the tannin itself, with pine tannin being more reactive than mimosa tannin. At the acid pH, the reaction is both reaction of tannin with furfuryl alcohol as well as self-condensation of furfuryl alcohol, whereas under rather alkaline conditions furfuryl alcohol cannot really self-condense and thus is forced to react with the tannin. It must be pointed out that the reactivity of the tannin even with aldehydes progressively increases from pH 4 (minimum reactivity) towards a more acidic pH, with the tannin being progressively more reactive as the pH become progressively lower. The same is true under alkaline conditions where the reactivity of the tannin increases and gel time decreases as one progresses to higher pH. The resins were prepared by mixing 100 parts of tannin with 100 parts of water and reacting this with 50 and 75 parts of furfuryl alcohol. The results were monitored by gel time measurements and thermomechanical analysis (TMA). The laboratory particleboard bonded with this resin under standard laboratory conditions and the dry internal bond (IB) strength was tested according to European Norm EN 312, 1995. The results confirmed that tannin extracts and furfuryl alcohol react with each other and do cross-link in the total absence of formaldehyde (**Table 2**).

For the mimosa tannin-furfuryl alcohol particleboards, the dry IB strength satisfies only marginally the requirements of the relevant European norm. However, for the pine tannin-furfuryl alcohol particleboards the dry IB strength satisfies the relevant EN 312 requirements.

Ten-ply and twenty-ply high continuous-type pressure paper laminates were prepared by impregnating filter paper with a mimosa-tannin solution mixed with

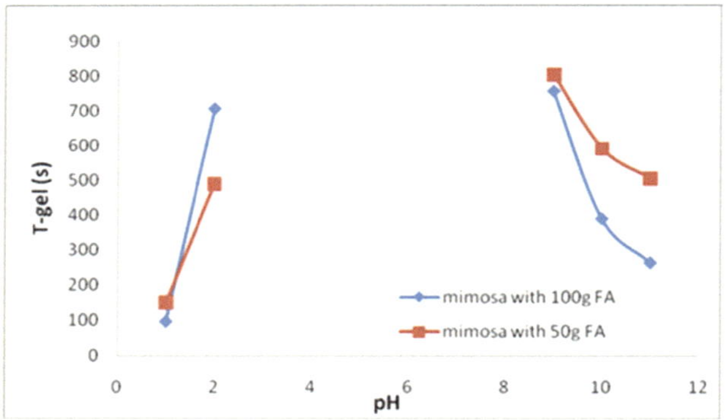

Figure 7.
Gel time of mimosa tannin reacted with furfuryl alcohol at pH ranging from 1 to 11. The gel time at pH 3, 4, 5, and 8 cannot be attained [63].

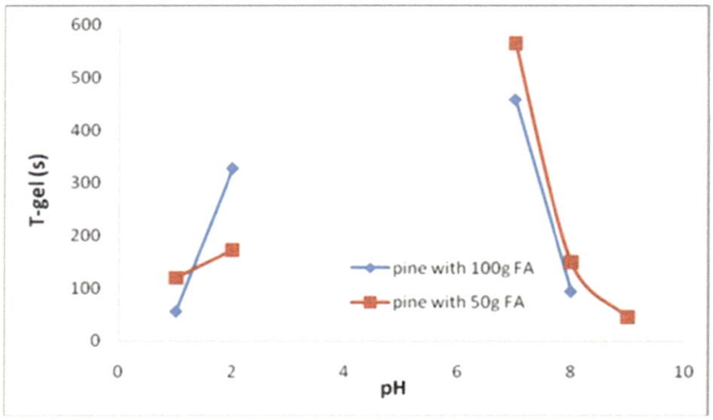

Figure 8.
Gel time of pine tannin reacted with furfuryl alcohol at pH ranging from 1 to 9. The gel time at pH 3, 4, and 5 cannot be attained [63].

	Gel time (s)	TMA max MOE (MPa)	Board density (kg/m^3)	IB strength (MPa)
Mimosa tan + 100%FA, pH 11	260	1929 ± 81	—	—
Mimosa tan + 50%FA, pH 11	500	2177 ± 82	—	—
Mimosa tan + 100%FA, pH 10	400	2332 ± 112	—	—
Mimosa tan + 50%FA, pH 10	600	2401 ± 97	716	0.34 ± 0.02
Pine tannin + 50% FA, pH 8	150	2430 ± 100	697	0.35 ± 0.02
Pine tannin + 75%FA, pH 8	110	3034 ± 130	715	0.40 ± 0.02

MOE = modulus of elasticity; IB = internal bond.

Table 2.
Results for wood particleboard panels bonded with furanic-tannin-based wood adhesives.

furfuryl alcohol and a formurea concentrate [64]. Crosscut, abrasion resistance, and water vapor resistance measurements were done. The effect of bonding 10-ply paper laminates on plywood shear strength was also determined. The 10-ply paper laminates with mimosa tannin-furfuryl alcohol resin appeared to increase the plywood dry shear strength while reducing its absorption of water. When pressed at 140°C temperature at 120 kg cm^2 pressure for 600 s, the 10-ply paper laminates gave the best appearance compared to other laminates.

The syntheses of difurfuryl diisocyanates [e.g., ethylidenebis (2,5-furandiylmethylene) diisocyanate (EDFI)] with formula shown in **Figure 9** have been reported in the literature [65].

Difurfuryl diisocyanates (**Figure 9**) are structurally similar to diphenylmethane diisocyanate (MDI), hence they can be equally good adhesives for bonding wood composites. The EDFI adhesive is synthesized from biomass-derived chemicals, contrary to the petroleum-derived MDI. The mechanical performances of MDI- and EDFI-bonded aspen flakeboards were compared. Flakeboards bonded with MDI showed results only marginally better than those bonded with EDFI. The difference has been ascribed to EDFI having greater viscosity than MDI. This has been thought to have caused a less optimal distribution of EDFI during spraying on the wood flakes, causing the slight difference in strength properties. The dry IB strength values of EDFI-bonded flakeboards showed dry IB strengths of 0.97 MPa, hence a value significantly greater than the 0.41 MPa required by the American National Standards Institute for type-2 medium-density particleboard when compared to MDI yielding 1.13 MPa.

Furfuryl alcohol, a biosourced material, is widely used in the foundry industry and in adhesives as additives or modifiers. However, furanic resins have not been reported as being used alone as wood panel adhesives. Furfuryl alcohol-aldehyde resins were nonetheless recently prepared for wood panel adhesives by reacting furfuryl alcohol with three different aldehydes: formaldehyde, glyoxal, and glutaraldehyde [66]. p-Toluene sulfonic acid coupled with an acid self-neutralizing system to minimize any damage to the wood substrate was used as a resin hardener to prepare plywood panels and to determine their bonding performances. In this adhesive system, formaldehyde and glyoxal reacted with furfuryl alcohol and the resin so prepared had excellent performance. The reaction of glutaraldehyde with furfuryl alcohol instead seemed difficult, the furfuryl

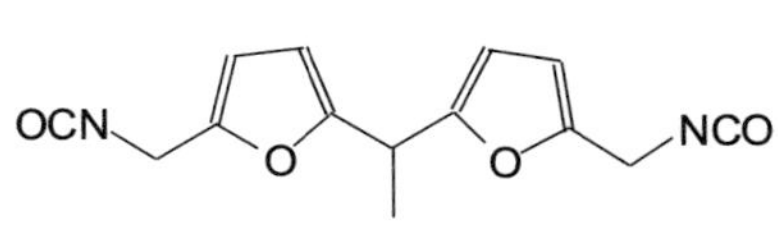

Figure 9.
Structural formula of difurfuryl diisocyanate.

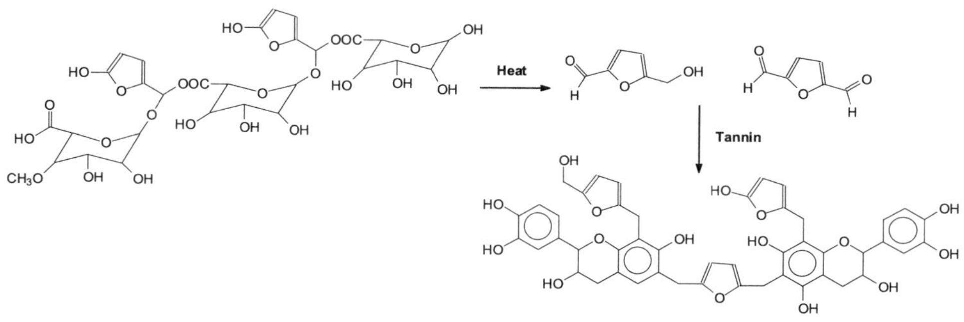

Figure 10.
Schematic representation of the formation of the reactive species hydroxymethyl furfural and furan 2,5-dialdehyde from carbohydrate exudates of several African trees and their reaction to cross-link tannins.

alcohol autocondensation predominating instead. The curing agent acidity greatly influences the resin bonding performance. The furfuryl alcohol-glyoxal resin showed a good bonding strength and water resistance greater than the standard requirements (≥0.7 MPa), even when using an acid self-neutralizing system as a hardener. This resin performed particularly well, considering that no formaldehyde was used.

Furfural and more recently hydroxymethyl furfural (HMF) are well-known upgraders of formaldehyde-based synthetic resins. An excellent review on this exists and the reader is addressed to it [67]. Even early literature and patents are known on this subject. The use of hydroxymethyl furfural is less known in synthetic resins where formaldehyde has been totally eliminated. Recent examples of the increased interest in HMF are the resins based on the coupling of glyoxal with HMF. Glyoxal is an aldehyde that is both nonvolatile and nontoxic. It can be used to substitute formaldehyde to prepare melamine-glyoxal (MG) resins for the wood industry. Due to the lower reactivity of glyoxal compared to formaldehyde, the MG resins performance is not as good as could be expected. Thus, 5-hydroxymethyl furfural (HMF) was used as a modifier to improve the properties of MG resins to prepare a hydroxymethyl furfural modified melamine-glyoxal (HMFMG) adhesive for bonding plywood [68]. The structure of the oligomers formed was determined along with the thermomechanical properties of the resins. The HMFMG resin presented a lower curing activation energy than the MG resin, yielding a much better bonded and water-resistant plywood.

Some fully biobased carbohydrate extracts from African trees have shown to release both hydroxymethyl furfural and furan 2,5-dialdehyde as hardeners [69, 70] during hot-pressing. **Figure 10** shows an example of the reactions involved.

This approach also fits with the adhesives based on the reaction of the reactive procyanidins of pine bark tannin with hydroxymethyl furfural [69] also yielding encouraging wood bonding results.

5. Furanic-based hard thermoset plastics

A 100% biosourced thermoset material based on condensed tannin-furfuryl alcohol thermoset resins has been used as the resin matrix of solid abrasive wheels by using pTSA as a catalyst [62, 71–73]. The system is based on two reactions: the reaction of furfuryl alcohol with the tannin and the acid-induced self-condensation of the furfuryl alcohol (**Figure 11**). The co-polymerization reactions were studied by ^{13}C NMR and MALDI-ToF mass spectrometry; they are shown in **Figure 11**.

The 100% renewable bioresourced tannin–furanic thermosetting resin was found to have a glass transition temperature as high as 211°C, and a 95% weight loss temperature of 244°C and 240°C in nitrogen and in air atmosphere, respectively.

The char yield is as high as 52%. Moreover, this new thermoset material showed excellent mechanical properties: a Brinell hardness of 23 HBS, which is higher than commercial acrylic, polyvinyl chloride and a little lower than that of solid (not foamed) polystyrene. The compressive break strength was found to be as high as 194.4 MPa, thus higher than that of filled phenolic resins and much higher than that of solid polystyrene and acetal resins (**Table 3**). **Figure 12** shows its appearance and the repetition of the resin stress vs. strain curves.

This hard thermoset resin was produced by a simple process that is easily industrialized. Abrasive wheels held together with this resin bonding different mineral and organic abrasive powders were developed and characterized. The main abrasive powder used was aluminum trioxide Al_2O_3 of different grit levels (**Figure 13**). Hard nutshell powders were also tried but did not give sufficiently good results. These abrasive wheels showed excellent abrasiveness properties when compared to commercial abrasive wheels [71].

Angle-grinder's cutting and grinding discs based on this green resin were also used for bonding abrasive particles of aluminium trioxide of different sizes and of different grits level (**Figure 13**). These discs were characterized and showed excellent abrasiveness and cutting properties. Their mechanical resistance was found to be comparable to that of commercial grinding discs bonded with synthetic phenolic resins. They tolerated well the severe stresses induced on them at 11,000 revolutions per minute (rpm) by operation in an angle grinder when grinding or cutting steel [72].

The same hard resin was used as a resin matrix for automotive brake pads. These experimental automotive brake pads based on this green resin showed excellent braking properties and wear resistance when used in a real car under full-scale test conditions. Their mechanical resistance was found to be comparable to that of commercial automotive brake pads bonded with synthetic phenolic resins. They tolerated well the severe stresses induced by strong braking, such as emergency braking at 50 km/h (31 mph) until complete standstill and showed braking distances comparable or even shorter than commercial brake pads [73].

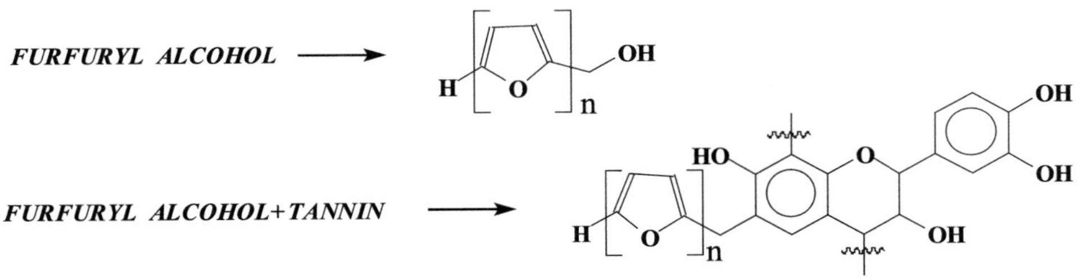

Figure 11.
Schematic representation of the reaction of furfuryl alcohol with tannin to form hard thermoset plastics.

Resin name	Breaking strength (MPa)	Young's modulus (GPa)
Tannin–furanic plastic	194.4 + 2.3	2.16 + 0.09
Solid Polystyrene	106	3.88
Acetal resin	100	3.28
Filled phenolic resin	158	6.82

[a]*Note: the values of the other resins are cited from ASTM_D695-10.*

Table 3.
Breaking strength and Young's modulus of tannin–furanic resin under compression.[a]

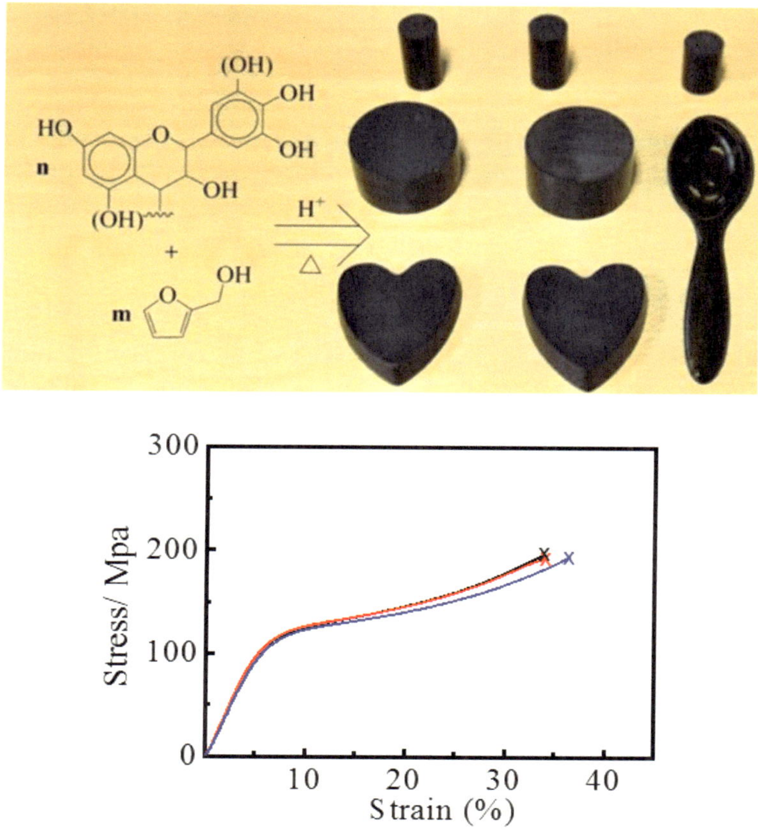

Figure 12.
Top: examples of rigid plastic specimens prepared by the reaction of tannin and furfuryl alcohol. Bottom: plot of the stress/strain curves of the tannin-furfuryl rigid plastic, the two curves showing its behavior repeatability.

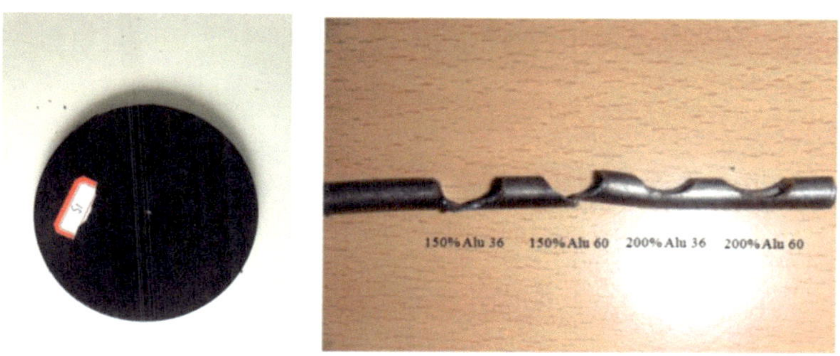

Figure 13.
Left: example of angle grinder disc formed by a tannin-furfuryl rigid plastic matrix and aluminium oxide abrasive. Right, example of a steel tube cut with the same type of disc but with different abrasive grits [72].

The same technology led to the preparation of both highly flexible films and strongly adhering non-scratch surface finishes by reacting partially aminated polyflavonoid tannins with furfuryl alcohol in the presence of plasticizers such as glycerol or polyethyleneimine. Chemical analysis showed partial amination of the tannin under the conditions used and even the formation of some –N= bridges between flavonoid units, although these were shown to be rare. Oligomers formed by the reaction of furfuryl alcohol with the flavonoid units and the simultaneous self-condensation of furfuryl alcohol were detected. Linear methylene–furanic chains were also found to be linked to flavonoid reactive sites. Side condensation reactions of furfuryl alcohol led to the formation of methylene ether bridges between furanic nuclei, followed by rearrangement to methylene bridges with liberation of formaldehyde. The latter reacted with both the flavonoid units and furan ring reactive sites to yield $-CH_2OH$, $-CH_2^+$ groups and methylene bridges [74].

6. Conclusion

Furanic resins either alone or in combination with other renewable biosourced materials have come of age in fields where they were never considered before, either because of their dark color or high cost. Their biosourced, renewable materials-derived label has changed this perception in a world looking for materials that are not oil-derived. Thus, from their traditional industrial applications, namely, in foundry sand shell molds for metal casting, they are starting to be used other areas such as fire-resistant, thermally insulating, and sound absorbent foams, as wood panel adhesives, hard rigid plastics, resistant matrix resins for abrasive aggregates, and even for the formation of flexible films and non-scratch surface finishes. These biomaterials are attracting the interest of researchers to develop new uses for them. Thus, their future expansion to a variety of products appears now to be assured.

Acknowledgements

The LERMAB is supported by a grant of France's Agence Nationale de la Recherche (ANR) in the ambit of the laboratory of excellence (LABEX) ARBRE.

Conflict of interest

The authors declare no conflict of interests.

Author details

Antonio Pizzi[1]* and Anish Khan[2]

1 LERMAB-ENSTIB, University of Lorraine, Epinal, France

2 Department of Chemistry, King Abdulaziz University, Jeddah, Saudi Arabia

*Address all correspondence to: antonio.pizzi@univ-lorraine.fr

References

[1] Yahama Y, Slono T. Japan Patent JP36,003,395. 1964

[2] International Research and Development Corp. UK Patent GB 971,217

[3] Choura M, Belgacem N, Gandini A. Acid-catalyzed polycondensation of furfuryl alcohol: Mechanisms of chromophore formation and cross-linking. Macromolecules. 1996;**29**(11): 3839-3850

[4] Dyno Industries. Procédé de préparation de noyaux et des moules de fonderie à partir d'olivine. French Patent 2123550. 1972

[5] Zennaro G. Foundry binder system with low emission of aromatic hydrocarbons. European Patent EP 1531018. 2005

[6] Tondi G, Pizzi A, Pasch H, Celzard A, Rode K. MALDI-TOF investigation of furanic polymer foams before and after carbonization: aromatic rearrangement and surviving furanic structure. European Polymer Journal. 2008;**44**:2938-2943

[7] Roux DG, Ferreira D, Hundt HKL, Malan E. Structure, stereochemistry, and reactivity of natural condensed tannins as basis for their extended industrial application. Journal of Applied Polymer Science. Applied Polymer Symposium. 1975;**28**:335-353

[8] Meikleham N, Pizzi A. Acid and alkali-setting tannin-based rigid foams. Journal of Applied Polymer Science. 1994;**53**:1547-1556

[9] Tondi G, Pizzi A. Tannin based rigid foams: Characterisation and modification. Industrial Crops and Products. 2009;**29**:356-363

[10] Tondi G, Pizzi A, Masson E, Celzard A. Analysis of gasses emitted during carbonization degradation of polyflavonoid tannin/furanic rigid foams. Polymer Degradation and Stability. 2008;**93**:1539-1543

[11] Li X, Basso MC, Fierro V, Pizzi A, Celzard A. Chemical modification of tannin/furanic rigid foams by isocyanates and polyurethanes. Maderas. 2012;**14**:257-265

[12] Tondi G, Zhao W, Pizzi A, Fierro V, Celzard A. Tannin-based rigid foams: A survey of chemical and physical properties. Bioresource Technology. 2009;**100**:5162-5169

[13] Tondi G, Blacher S, Leonard A, Pizzi A, Fierro V, Leban JM, et al. X-ray microtomography studies of tanninderived organic and carbon foams. Microscopy and Microanalysis. 2009;**15**:395-402

[14] Zhao W, Fierro V, Pizzi A, Du G, Celzard A. Effect of composition and processing parameters on the characteristics of tannin-based rigid foams. Part 2: Physical properties. Materials Chemistry and Physics. 2010;**123**:210-217

[15] Zhao W, Pizzi A, Fierro V, Du G, Celzard A. Effect of composition and processing parameters on the characteristics of tannin-based rigid foams. Part 1: Cell structure. Materials Chemistry and Physics. 2010;**122**: 175-182

[16] Celzard A, Fierro V, Amaral-Labat G, Pizzi A, Torero J. Flammability assessment of tannin-based cellular materials. Polymer Degradation and Stability. 2011;**96**: 477-482

[17] Lacoste C, Basso MC, Pizzi A, Celzard A, Ella Bang E, Gallon N, et al. Pine (*Pinus pinaster*) and quebracho (Schinopsis lorentzii/balansae) tannin

based foams as green acoustic absorbers. Industrial Crops and Products. 2015;**67**:70-73

[18] Chen X, Li J, Pizzi A, Fredon E, Gerardin C, Zhou X, et al. Tannin-furanic foams modified by soybean protein isolate (SPI) and lignin substituting formaldehyde addition. Industrial Crops and Products. 2021;**168**:113607. DOI: 10.1016/j.indcrop.2021.113607

[19] Basso MC, Pizzi A, Al-Marzouki F, Abdalla S. Horticultural/hydroponics and floral foams from tannins. Industrial Crops and Products. 2016;**87**:177-181

[20] Basso MC, Giovando S, Pizzi A, Pasch H, Pretorius N, Delmotte L. Flexible-elastic copolymerized polyurethane-tannin foams. Journal of Applied Polymer Science. 2014;**131**:13. DOI: 10.1002/app.40499

[21] Abdalla SMS, Al-Marzouki F, Pizzi A, Bahabri FS. Bone graft with a tannin-hydroxyapatite scaffold and stem cells for bone engineering. US Patent number. 2018;**10**: 155-069

[22] National Institute for Occupational Safety and Health (NIOSH). Registry of Toxic Effects of Chemical Substances. Washington, DC, USA: National Institute for Occupational Safety and Health; 2000

[23] National Technical Information Service (NTIS). Publication AD-A125-539. Washington, DC, USA: National Technical Information Service; 1977

[24] Basso MC, Li X, Giovando S, Fierro V, Pizzi A, Celzard A. Green, formaldehyde-free foams for thermal insulation. Advanced Materials Letters. 2011;**2**(6):378-382

[25] Pizzi A, Basso MC, Celzard A, Fierro V, Giovando S. Composition for manufacturing a tannin-based foam material, foam material obtainable from it, and manufacturing process thereof. US patent 9,302,413 (2016)

[26] Basso MC, Giovando S, Pizzi A, Celzard A, Fierro V. Tannin/furanic foams without blowing agents and formaldehyde. Industrial Crops and Products. 2013;**49**:17-22

[27] Basso MC, Pizzi A, Celzard A. Influence of formulation on the dynamics of preparation of tannin based foams. Industrial Crops and Products. 2013;**51**:396-400

[28] Li X, Pizzi A, Cangemi M, Fierro V, Celzard A. Flexible natural tannin-based and protein-based biosourced foams. Industrial Crops and Products. 2012;**37**:389-393

[29] Li X, Essawy H, Pizzi A, Delmotte L, Rode K, Le Nouen D, et al. Modification of tannin based rigid foams using oligomers of a hyperbranched poly(amine-ester). Journal of Polymer Research. 2012;**19**:21-29

[30] Lacoste C, Basso MC, Pizzi A, Laborie MP, Celzard A, Fierro V. Pine tannin-based rigid foams: mechanical and thermal properties. Industrial Crops and Products. 2013;**43**:245-250

[31] Lacoste C, Basso MC, Pizzi A, Laborie MP, Garcia D, Celzard A. Bioresourced pine tannin/furanic foams with glyoxal and glutaraldehyde. Industrial Crops and Products. 2013;**45**:401-405

[32] Lacoste C, Pizzi A, Basso MC, Laborie MP, Celzard A. Pinus pinaster tannin/furanic foams: Part 1, formulations. Industrial Crops and Products. 2014;**52**:450-456

[33] Lacoste C, Pizzi A, Basso MC, Laborie MP, Celzard A. Pinus pinaster tannin/furanic foams: Part 2: physical properties. Industrial Crops and Products. 2014;**61**:531-536

[34] Lacoste C, Cop M, Kampainen K, Giovando S, Pizzi A, Laborie MP, et al. Biobased foams from condensed tannins from Norway spruce (*Picea abies*) bark. Industrial Crops and Products. 2015;**73**:144-153

[35] Cop M, Laborie MP, Pizzi A, Sernek M. Curing characterisation of spruce tannin-based foams using the advances isoconversional method. BioResources. 2014;**9**(3):4643-4655

[36] Cop M, Lacoste C, Conradi M, Laborie MP, Pizzi, Sernek M. The effect of the composition of spruce and pine tannin-based foams on their physical, morphological and compression properties. Industrial Crops and Products. 2015;**74**:158-164

[37] Cop M, Gospodaric B, Kempainen K, Giovando S, Laborie MP, Pizzi A, et al. Characterization of the curing process of pine and spruce tannin-based foams by different methods. European Polymer Journal. 2015;**69**:29-37

[38] Rangel G, Chapuis H, Basso MC, Pizzi A, Delgado-Sanchez FV, Celzard A, et al. Improving water repellency and friability of tannin-furanic foams by oil-grafted flavonoid tannins. BioResources. 2016;**11**: 7754-7768

[39] Zhou X, Pizzi A, Sauget A, Nicollin A, Li X, Celzard A, et al. Lightweight tannin foam/composites sandwich panels and the coldset tannin adhesive to assemble them. Industrial Crops and Products. 2013;**43**:255-260

[40] Link M, Kolbitsch C, Tondi G, Ebner M, Wieland S, Petutschnigg A. Formaldehyde-free tannin based foams and their use as lightweight panels. BioResources. 2011;**6**(4):4218-4228

[41] Basso MC, Lagel MC, Pizzi A, Celzard A, Abdalla S. First tools for tannin-furanic foams design. BioResources. 2015;**10**(3):5233-5241

[42] Meikleham N, Pizzi A, Stephanou A. Induced accelerated autocondensation of polyflavonoid tannins for phenolic polycondensates, Part 1: 13C NMR, 29Si NMR, x-ray and polarimetry studies and mechanism. Journal of Applied Polymer Science. 1994;**54**:1827-1845

[43] Pizzi A, Meikleham N. Induced accelerated autocondensation of polyflavonoid tannins for phenolic polycondensates—Part III: CP-MAS 13C NMR of different tannins and models. Journal of Applied Polymer Science. 1995;**55**:1265-1269

[44] Pizzi A, Meikleham N, Dombo D, Roll W. Autocondensation-based, zero-emission, tannin adhesives for particleboard. Holz als Roh- und Werkstoff. 1995;**53**:201-204

[45] Garcia R, Pizzi A. Polycondensation and autocondensation networks in polyflavonoid tannins, Part 1: Final networks. Journal of Applied Polymer Science. 1998;**70**:1083-1091

[46] Garcia R, Pizzi A. Polycondensation and autocondensation networks in polyflavonoid tannins, Part 2: Polycondensation vs. autocondensation. Journal of Applied Polymer Science. 1998;**70**:1093-1110

[47] Szczurek A, Fierro V, Medjahdi G, Celzard A. Carbon aerogels prepared by autocondensation of flavonoid tannin. Carbon Resources Convention. 2019;**2**:72-84

[48] Tondi G, Link M, Kolbitsch C, Lesacher R, Petutschnigg A. Pilot plant up-scaling of tannin foams. Industrial Crops and Products. 2016;**79**:211-218

[49] Tondi G, Link M, Oo CW, Petutschnigg A. A simple approach to distinguish classic and formaldehyde-free tannin based rigid foams by ATR FT-IR. Journal of Spectroscopy. 2015: 1-8; article ID 902340

[50] Reyer A, Tondi G, Berger RJF, Petutschnigg A, Musso M. Raman spectroscopic investigation of tannin-furanic rigid foams. Vibrational Spectroscopy. 2016;**84**:58-66

[51] Kolbitsch C, Link M, Petutschnigg A, Wieland S, Tondi G. Microwave produced tannin-furanic foams. Journal of Materials Science Research. 2012;**1**(3):84-91

[52] Tondi G, Johansson M, Leijonmarck S, Trey S. Tannin based foams modified to be semi-conductive: Synthesis and characterization. Progress in Organic Coatings. 2015;**78**:488-493

[53] Tondi G, Link M, Kolbitsch C, Petutschnigg A. Infrared-catalyzed synthesis of tannin-furanic foams. BioResources. 2013;**9**(1):984-993

[54] Basso MC, Pizzi A, Lacoste C, Delmotte L, Al-Marzouki FA, Abdalla S, et al. Tannin-furanic-polyurethane foams for industrial continuous plant lines. Polymers. 2014;**6**:2985-3004

[55] Garcia DE, Fuentealba CA, Salazar JP, Perez MA, Escobar D, Pizzi A. Mild hydroxypropylation of polyflavonoids obtained under pilot-plant scale. Industrial Crops and Products. 2016;**87**:350-362

[56] Garcia DE, Glasser WG, Pizzi A, Paczkowski S, Laborie MP. Hydroxypropyl tannin from Pinus pinaster bark as polyol source in urethane chemistry. European Polymer Journal. 2015;**67**:152-165

[57] Pizzi A. Advanced Wood Adhesives Technology. New York: Marcel Dekker; 1994. p. 289

[58] Pizzi A, Walton T. Non-emulsifiable, water-based diisocyanate adhesives for exterior plywood, Part 1: Novel reaction mechanisms and their chemical evidence. Holzforschung. 1992;**46**:541-547

[59] Pizzi A, Valenzuela J, Westermeyer C. Non-emulsifiables, water-based, diisocyanate adhesives for exterior plywood, Part 2: industrial application. Holzforschung. 1993;**47**: 69-72

[60] Nicollin A, Li X, Girods P, Pizzi A, Rogaume Y. Fast pressing composite using tannin-furfuryl alcohol resin and vegetal fibers reinforcement. Journal of Renewable Materials. 2013;**1**(4):311-316

[61] Sauget A, Nicollin A, Pizzi A. Fabrication and mechanical analysis of mimosa tannin and commercial flax fibers biocomposites. Journal of Adhesion Science and Technology. 2013;**27**:2204-2218

[62] Li X, Nicollin A, Pizzi A, Zhou X, Sauget A, Delmotte L. Natural tannin/ furanic thermosetting moulding plastics. RSC (Royal Chemical Society) Advances. 2013;**3**:17732-17740

[63] Abdullah UHB, Pizzi A. Tannin-Furfuryl alcohol wood panel adhesives without formaldehyde. European Journal of Wood and Wood Products. 2013;**71**(1):131-132

[64] Abdullah UHB, Pizzi A, Zhou X. High pressure paper laminates from mimosa tannin resin. International Wood Products Journal. 2014;**5**(4): 224-227

[65] Holfinger MS, Conner AH, Lorenz LF, Hill CG Jr. Difurfuryl diisocyanates: New adhesives derived from renewable resources. Journal of Applied Polymer Science. 1993;**49**(2): 337-344

[66] Xi X, Wu Z, Pizzi A, Gerardin C, Lei H, Du G. Furfuryl alcohol-aldehyde plywood adhesive resins. The Journal of Adhesion. 2020;**96**(9):814-838

[67] Belgacem MN, Gandini A. Furan-based adhesives. In: Pizzi A, Mittal KM,

editors. Handbook of Adhesive Technology. 2nd ed. New York: Marcel Dekker; 2003. pp. 615-634

[68] Xi X, Liao J, Pizzi A, Gerardin C, Amirou S, Delmotte L. 5-hydroxymethyl furfural modified Melamine Glyoxal resin. The Journal of Adhesion. 2020;**96**(13):1167-1185

[69] Ndiwe B, Pizzi A, Tibi B, Danwe R, Konai N, Amirou S. African tree bark exudate extracts as biohardeners of fully biosourced thermoset tannin adhesives for wood panels. Industrial Crops and Products. 2019;**132**:253-268

[70] Ndiwe B, Pizzi A, Tibi B, Danwe R, Konai N, Amirou S. Particleboard bonded with biohardeners of tannin adhesives. European Journal of Wood and Wood Products. 2019;**77**(6): 1221-1223

[71] Lagel MC, Pizzi A, Basso MC, Abdalla S. Development and characterisation of abrasive grinding wheels with a tannin-furanic resin matrix. Industrial Crops and Products. 2014;**65**(3):333-348

[72] Lagel MC, Zhang J, Pizzi A. Cutting and grinding wheels for angle grinders with a bioresin matrix. Industrial Crops and Products. 2015;**67**:264-269

[73] Lagel MC, Hai L, Pizzi A, Basso MC, Delmotte L, Abdalla S, et al. Automotive brake pads made with a bioresin matrix. Industrial Crops and Products. 2016;**85**(3):372-381

[74] Basso MC, Lacoste C, Pizzi A, Fredon E, Delmotte L. Flexible tannin-furanic films and lacquers. Industrial Crops and Products. 2014;**61**:352-360

Chapter 7

Furfural: A Versatile Derivative of Furan for the Synthesis of Various Useful Chemicals

Kazeem Adelani Alabi, Rasheed Adewale Adigun, Ibrahim Olasegun Abdulsalami and Mariam Dasola Adeoye

Abstract

Furfural, a five-membered heterocyclic aromatic hydrocarbon derivable from acid hydrolysis of sugar cane bagasse, maize cob, rice husk or any cellulose-containing material, is useful in the synthesis of a range of specialized chemical products. Its condensation with nitromethane in basic medium yields 2-(2-Nitrovinyl) furan. This functional group (nitrovinyl) has been documented as a potent anti-microbial agent against gram-positive and gram-negative bacteria, with more potency against the gram-positive strains. The reaction of urea and thiourea with furfural yields bisimines-1,3-bis[(*E*)-furan2-yl)methylene]urea, and 1,3-bis[(*E*)-furan-2-yl) methylene]thiourea respectively. The two compounds are good antimicrobial agents in addition to the latter as a potential dye for wool and cotton fabrics with different hues. Also the reaction between acetophenone and furfural (an aldehyde) in a basic medium yields the chalcone: (*E*)-3-(furan-2-yl)-1-phenylprop-2-ene-1-one. This chalcone has been confirmed as a good antifungal agent and wood-protector against termite attack. Thus, chemical modification of the aldehyde functional group of furfural to nitro, imine and chalcone groups imparted different activities on furfural.

Keywords: Furfural, Furan, 2-(2-Nitrovinyl) furan, Bisimine, Chalcone, Extraction

1. Introduction

Furfural is a utility chemical. It is very simple to manufacture, the raw material is always agricultural waste, and the cost of production is reasonably low and affordable. It is fondly called "Gold from Garbage" [1]. Mostly, undiluted hydrogen tetraoxosulphate (VI) acid is used for the extraction of furfural from biomass that contain pentosans, which are aldose sugars, that composed of small rings formed from short five-member chains, that constitute a class of complex carbohydrates, present in the cellulose of many woody plants such as bagasse (sugar cane waste) corn cobs, rice, and oat hulls, etc. [2].

Furfural is a heterocyclic aldehyde with oxygen as the heteroatom replacing one of the carbon atoms of the five membered ring. It has a density of 1160 kg/m^3 and boils at 161.7°C. It is though, colorless, but changes to dark brown when it reacts with the atmospheric oxygen. It has other names apart from furfural; furaldehyde, 2-furanaldehyde, and 2-carbaxylaldehyde [3].

The flexible production route of furfural makes it to be the most commonly produced chemical in the industry, in addition to its many industrial uses and renewability. It is one of the top value-added chemicals that is being produced from biomass. Furfural and its derivatives have been extensively used in plastics, pharmaceutical, agrochemical and it has been reported to be used as dye precursors in the synthesis of diazo disperse dyes [4].

Sometimes, it is used as a solvent or starting material for the production of various chemicals. In order to increase its production because of its usefulness, furfural is now synthesized in the laboratory through different sources [2].

The process of furfural production is well understood but it's limited by low yield, very high energy consumption, and environmental and safety risks attached to the corrosive homogenous acid catalysts being used [5]. Hence, the existing method is being modified to make it green by the use of heterogeneous catalysts. Westpro-modified Huaxia Technology and Supra Yield are two improved processes where high furfural yields, recovery of the catalysts and high purity are obtainable [6]. Hybrid solid acid catalysts are recently used to achieve better dehydration of xylose to form furfural [4]. Furfuryl alcohol has been used for the production of furan resin prepolymer material which is useful for the production of different thermoset polymers [4].

2. Reactivity of furfural

Furfural undergoes some chemical reactions to yield different products. Among the reactions are hydrogenation, oxidation, decarboxylation/decarbonylation and condensation reactions.

2.1 Hydrogenation of furfural

The hydrogenation of furfural is catalyst dependent. The carbonyl functional group (C=O) or the aromatic ring will be reduced [7]. Almost all hydrogenation catalysts are capable of hydrogenating the carbon–carbon double bonds (C=C) and the carbonyl group (C=O) of furfural. The control of the hydrogenation process of these bonds has been one of the most important subjects in the developmental study of catalysts [8]. The addition of hydrogen to the C=C bond is more thermodynamically and kinetically favorable compared to the C=O bond. There is a need for a chemoselective hydrogenation catalyst within a standard conditions [9]. Hydrogenation reaction takes place in liquid and gaseous phases. Despite the high yields reported for the liquid phase, industrial hydrogenation of furfural is mostly carried out in the gas phase because of its lower cost as compared to the liquid phase (**Figure 1**) [10].

2.2 Oxidation of furfural

Furfural is easily oxidized even with atmospheric air to 2-furoic acid which is a heterocyclic carboxylic acid (**Figure 2**). Its name is derived from the Latin word furfur, meaning bran. The salts of this acid is called furoate. Many scientists/

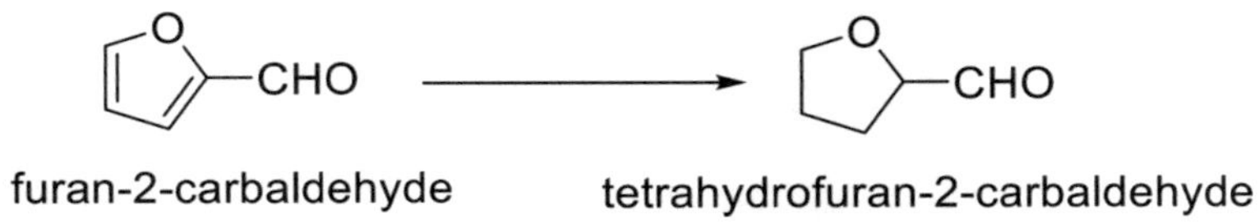

Figure 1.
Hydrogenation of furfural.

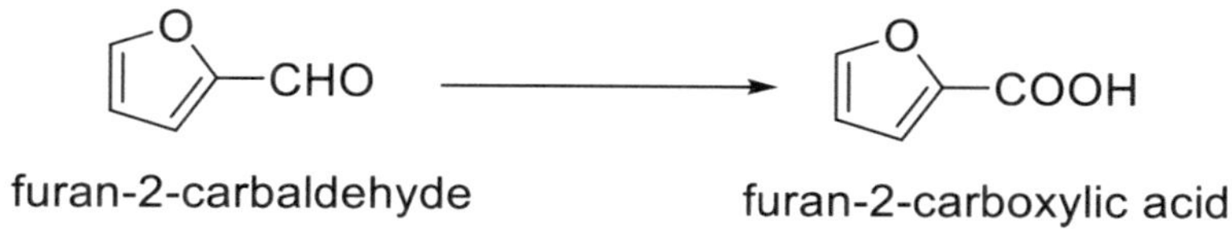

Figure 2.
Oxidation of furfural to furoic acid.

researchers have worked in this area (oxidation of furfural). In addition, oxidative condensation of furfural with alcohol has produced biofuel useful in the aviation industry [11].

Gold supported catalyst condensed furfural with ethanol in the presence of oxygen molecule under mild conditions [12]. Au/ZrO_2 catalysts was also used in another reaction to produce furoate ester [13]. Oxidative esterification of furfural with oxygen and methanol was also performed [14]. The effect of catalyst was evaluated in oxidative esterification of furfural [15]. The furfural can also be oxidized to produce maleic acid in the presence of hydrogen peroxide and a solid catalyst. The maleic acid can be dehydrated to maleic anhydride. These compounds have many applications in the industry. They are used in the manufacture of polymers, resins, pharmaceuticals, and biofuels. The oxidation of furfural into maleic acid involves two steps; decarbonylation and oxidation.

2.3 Decarboxylation/decarbonylation reaction

Two methods are well known to transform furfural into furan. When furfural is passed over metal surface in the complete absence of air and water at a very high temperature; furan and carbonmonoxide are obtained. The other process involves passing furfural together with excess steam over chromite or soda-lime catalyst, in the presence of hydrogen but absence of oxygen. Furan, carbondioxide and hydrogen are the resulted products. Alternatively, furfural may be converted to furoic acid at 300–500°C, on the surface of any higher oxide of heavy metals in the presence of oxygen, and the acid is immediately decarboxylated to give furan.

Generally, furoic acid is the product of oxidation of furfural. Decarboxylation of this acid or decarbonylation of furfural with the help of catalysts is the prevailing way for furan production in both gas and liquid phases [16]. Furan is costlier than furfural. In industry, the gas phase method is more attractive because of its advantages of simple operation and the easy separation and recycling of catalysts. Due to the high energy barrier for the C–C bond cleavage, the decarbonylation of furfural is preferably conducted at high temperatures. Additionally, hydrogen is always present in this reaction to maintain the metallic state of catalysts. Furfural is highly stable because of its high resonance energy [16]. Decarbonylation of furfural with high efficiency still presents considerable challenges. Among the various catalysts employing different metals and supports, Pd catalysts loaded on carbon, alumina and silica have demonstrated superior activity due to the favorable formation of acyl surface species, the key adsorption intermediate for decarbonylation [16]. Recently, Platinum has attracted much attention due to its cheapness and good activity [16].

2.4 Condensation reaction

Furfural condenses with nitromethane in the presence of a base to yield Nitrovinyl furan derivative [2]. Any compound containing active alpha-hydrogen in the presence of a base (ammonia or amines) as a catalyst will condense with furfural according to Knoevenagel condensation [2].

The aldol condensation of acetone and furfural has been deeply studied using different homogeneous and heterogeneous catalysts, highlighting different hydroxides (NaOH and $Ca(OH)_2$) and mixed oxides such as Mg-Zr or Mg-Al [17].

Cyclopentanone can undergo condensation with furfural or 2-methyl furan through an alkylation reaction to produce oxygen-containing compounds, which can in-turn, be converted into high-density diesel or jet fuel by hydro deoxygenation. Aldol condensation of cyclopentanone and furfural by sodium hydroxide in a solvent-free system produced a high yield of 2,5-bis (2-furylmethylidene) cyclopentanone [18].

3. Imine

Imine or azo-methine is derivable from nature and also synthesized in the laboratory. The azo-methine functional group (C=N) is responsible for a number of biological activities [19]. Apart from the bioactivities, there are enough reports in the literature that azo-methine also exhibits pigments and dyes properties, used as catalysts, and as stabilizers in polymeric products [20]. Azo-methine also possesses herbicidal activities [21]. The formation of metal-complexes with imine as ligand improves its uses [22].

Complexes formed by reacting imine with transition metal have numerous applications. They have biological, industrial, catalystic applications [23]. These compounds are very important pharmaceutical precursors for their wide spectrum of biological activities [24].

They are very much applicable in dyes industries also; they perfectly dye leathers, food packages and wools with different hues [25].

Copper complexes are the most potent transition metal-imine complexes ever formed. They are very useful in chemical sciences like the food industry, dye industry, analytical chemistry, catalysis and many other fields [26].

The complexes formed with zinc metal exhibit good photo luminescent and electroluminescent activities. These compounds show good color purity and could be used in the production of full-colored organic light-emitting diodes [27]. They could also be used as dopant materials [28].

4. Synthesis and potential uses of 1,3-bis((E)-furan2-yl)methylene)urea (UBI)

In our recent study furfural was redistilled, separately mixed with urea and thiourea in the presence of glacial acetic acid (drops) under reflux to obtain

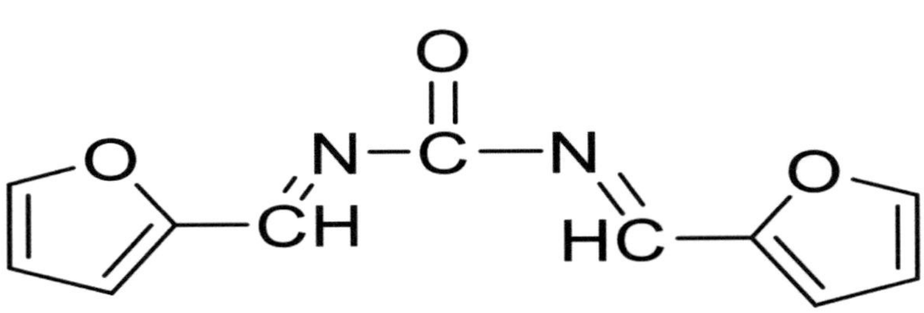

Figure 3.
Structure of 1,3-bis((E)-furan-2-ylmethylene)urea.

1,3-bis((E)-furan2-yl)methylene)urea and 1,3-bis((E)-furan2-yl)methylene) thiourea respectively. The products were filtered, washed with water and recrystallized in alcohol (**Figure 3**) [4].

A study has shown that 1,3-bis(furan-2-yl) methylene) urea (UBI) possesses hepatotoxic and nephrotoxic effects [19]. UBI has been reported to be a potential inhibitor of *P. aeruginosa* and *S. bovis* [4].

5. Potential uses of 1,3-bis((E)-furan-2-yl) methylene) thiourea (TBI)

Loomis and Hayes stated that any substance that causes harm to an animal with a dose below 5 mg/ kg could be regarded as a highly toxic substance [29]. Therefore, 1,3-bis((E)-furan-2-yl) methylene) thiourea cannot be regarded as being toxic since it did not cause harm to any of the mice administered with doses of 50 and 150 mg/kg.

TBI has been referred to as multifunctional dye when applied on wool and cotton fabrics. Because of its very high and moderate inhibitions rate on fungi and bacteria. The dyed fabrics can be utilized in the production of medical face masks. Excellent Ultraviolet Protection Factor (UPF) of the dyed fabrics with TBI is an indicator of its potency as a sun-screen protector, which may protect the users from harmful rays of sun burn. Its high fastness enhances the use of the dyed fabric for a very long time without fading. Also, it was reported that its mechanical properties make it a potential materials for engineering laboratory coats production [29].

6. Toxicity and the potential uses of 2-(2-nitrovinyl) furan

2-(2-Nitrovinyl)furan is a lipid-soluble substance that has been discovered to have antimicrobial properties, against skin-related microbial infections [30].

The compound 2-(2-Nitrovinyl)furan significantly reduced alkaline phosphatase, alanine, and aspartate aminotransferase activities in male rat liver and kidney with a corresponding increase in serum. The activities of superoxide dismutase, catalase, glutathione peroxidase, glutathione reductase, and levels of reduced glutathione/glutathione disulfide (GSSG) ratio in the liver and kidney of 2-(2-nitrovinyl)furan-treated rats decreased significantly. In contrast, GSSG, protein carbonyl, conjugated dienes, lipid hydroperoxides, malondialdehyde, and fragmented DNA (%) in 2-(2-nitrovinyl)furan-treated rats increased significantly. 2-(2-nitrovinyl) furan exhibited its toxicity by decreasing the antioxidant systems [31].

2-(2-nitrovinyl) furan exhibited additive interactions with chloramphenicol, erythromycin, lincomycin and gemifloxacin [32].

It has been reported that malondialdehyde and fragmented DNA level increased significantly in the bacterial cells treated with 2-(2-nitrovinyl) furan when compared with dimethylsulfoxide (DMSO) treated cells. The colony-forming unit (CFU) of *Escherichia coli*, *Pseudomonas aeruginosa* and *Staphylococcus aureus* following exposure to 2-(2-nitrovinyl)furan increased significantly ($p < 0.05$) in the presence of 2,2′ bipyridyl, an iron chelator, when compared with only 2-(2-nitrovinyl)furan suggesting the involvement of hydroxyl radical in the cell death. This study showed that 2-(2-nitrovinyl)furan induced oxidative stress in *Escherichia coli*, *Pseudomonas aeruginosa* and *Staphylococcus aureus* as evident from elevated levels of superoxide anion radical nitric oxides and antioxidant enzymes (**Figure 4**).

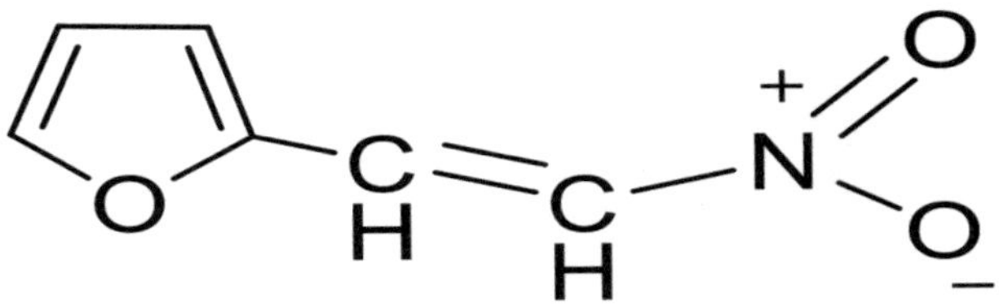

Figure 4.
Structure of 2-(2-nitrovinyl) furan.

7. Natural chalcone, extraction and uses

The name "Chalcones" was given by [33]. Chalcone is a generic term used in describing compounds with the 1, 3- diphenylprop-2-en-1- one frame work (**Figure 5**).

They are natural compounds that are largely found in plants, fruits and vegetables [34]. They are mostly found in various plants species like Angelica, Glycyrrhiza, Humulus and Scutellaria, which are widely used as traditional remedies. Most of the aromatic rings of natural chalcones are found in the hydroxylated form. Chalcones, dihydrochalcones and aurones are composed of pigments whose color changes from yellow to orange in some plant species. Those compounds are found not only in flowers but also in lots of different tissues of plants. Free radical scavenging properties of phenol groups of chalcones increase the interest in consumption of the plants that contain chalcones. The chemistry of chalcones has generated intensive scientific studies throughout the world.

Chemically they consist of open chain flavonoids in which two aromatic rings are joined by a three-carbon α, β carbonyl system. To develop most of the biological and pharmacoligical agents, a series of chalcones were prepared by Claisen-Schmidt condensation of appropriate acetophenones with appropriate aromatic aldehydes in the presence of an aqueous solution of potassium hydroxide and ethanol at room temperature [35]. Chalcones either natural or synthetic are known to exhibit various biological and pharmacological activities.

Chalcone is an aromatic ketone and enone that forms the central core for a variety of important biological compounds [36].

8. Synthesis and potential uses of furfural based chalcone

In one of our studies, (E)-3-(furan-2-yl)-1-phenylprop-2-en-1-one was synthesized using Claisen-Schmidt condensation. Sodium hydroxide (3 g) was dissolved in 10 ml of methanol. Acetophenone (3 mL) and 3 mL furfural were dissolved in 10 ml of rectified ethanol in a 250 ml conical flask on a magnetic stirrer. Then 10 ml NaOH solution (3 g in 10 ml methanol) was added drop wise to the reaction mixture on vigorous stirring for 30 minutes until the solution became turbid. The reaction temperature was maintained between 20 and 25°C using a cold water bath on the magnetic stirrer. After vigorous stirring for 4–5 hours the mixture started forming precipitate. The reaction vessel was kept overnight in the refrigerator to give room for complete precipitation. The precipitate formed was filtered, air-dried and recrystallized in rectified ethanol to give the pure product (**Figure 6**) which was weighed and stored for further analysis.

The presence of a reactive α, β unsaturated ketone functionality in chalcones is found to be responsible for these activities. In recent years, many studies have been reviewed and chalcone is found to be useful for their cytotoxic, anti-cancer,

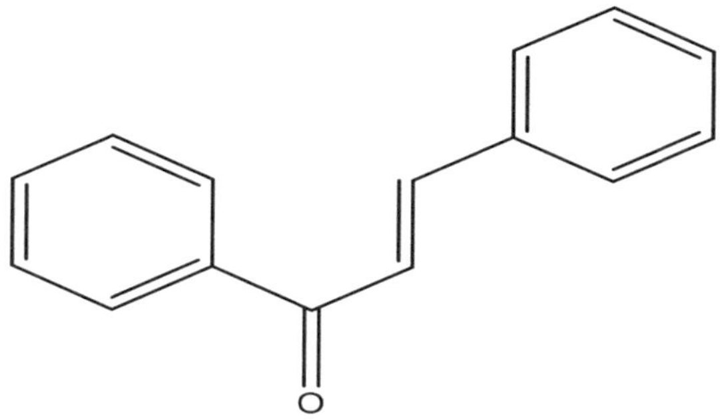

Figure 5.
The generic structure of chalcone.

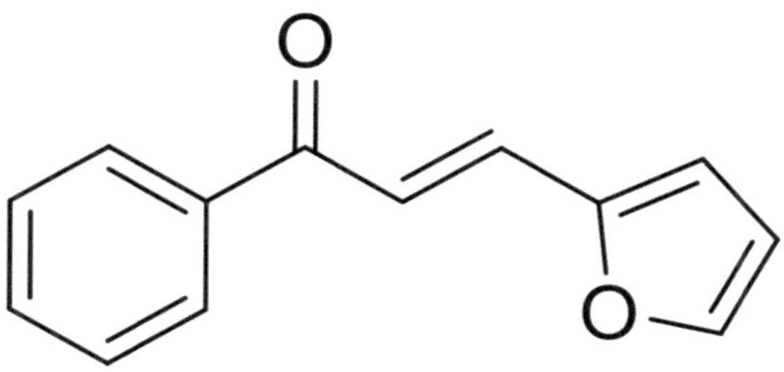

Figure 6.
Structure of (E)-3-(furan-2-yl)-1-phenylpro-2-en-1one.

anti-viral, anti-malarial, anti-oxidant, anti-leishmanial, anti-inflammatory, analgesic, anti-platelet, anti-ulcerative, anti-tubercular, anti-hyperglycemic, chemo-preventive, mutagenic, insecticidal and enzyme inhibitory properties [37].

9. Qualitative structure: activity relationship

The structures of the molecules have been modeled and geometry optimization as well as molecular orbital at DFT/B3LYP/6-31G* level of theory in vacuum. The optimized geometries of the studied molecules were then used to obtain the ground state molecular geometry parameters. The highest occupied molecular orbital (HOMO) and the lowest unoccupied molecular orbital (LUMO) energy values obtained were used to calculate other parameters [4].

10. Chemical-activities relationship result

Table 1 presents Highest occupied molecular orbital (HOMO) and lowest unoccupied molecular orbital (LUMO) of furfural (starting material), 1,3-bis[(*E*)-furan2-yl)methylene]urea (UBI), and 1,3-bis[(*E*)-furan-2-yl) methylene]thiourea (TBI), 2-(2-Nitrovinyl) furan (NVF) and (*E*)-3-(furan-2-yl)-1-phenylprop-2-ene-1-one (FPP) (last two rows). Other parameters/properties were derived from these values using different formulas for each property. The energy of the HOMO is directly related to the ionization potential while that of the LUMO is related to the electron affinity [38]. The high value of HOMO in any molecule is an indication of its high tendency to donate electrons to molecules with empty and low-lying energy orbital acting as an acceptor.

Properties	Formulae	Furfural	UBI	TBI	NVF	FPP
Energy gap (eV)	$E_L - E_H$	10.17	4.54	3.50	3.89	2.45
Chemical potential (K)	$\frac{1}{2}(E_H - E_L)$	−5.09	−2.27	−1.75	−1.95	−1.23
Absolute Hardness (η)	$\frac{1}{2}(E_L - E_H)$	5.09	2.27	1.75	1.95	1.23
Softness (S)	$\frac{1}{2\eta}$	0.10	0.22	0.29	0.26	0.41
Electrophilicity	$\frac{K^2}{2\eta}$	2.55	1.14	0.88	0.99	0.62
Electronegativity	$\frac{I + EA}{2}$	9.29	4.38	4.12	4.57	4.25
Ionization potential, I (eV)	$-E_H$	9.73	6.65	5.87	6.51	3.03
Electron affinity, EA (eV)	$-E_L$	−0.44	2.11	2.37	2.62	5.48
Dipole moment, μ (D)		2.68	3.41	3.34	6.44	
Polarizability (α)			57.59	58.65	51.15	
E_{HOMO} (eV)		−9.73	−6.65	−5.87	−6.51	−3.03
E_{LUMO} (eV)		0.44	−2.11	−2.37	−2.62	−5.48

Table 1.
Global parameters.

The energy band gap of furfural the starting material was the highest (10.17 eV), followed by UBI (4.54 eV), then NVF (3.89 eV), TBI (3.50 eV) and the lowest was FPP (2.45 eV). Large HOMO–LUMO energy gap is an indication of high stability and consequently less activity for the molecule in chemical reactions [39]. Thus, furfural is the most stable with less activity and FPP is the least stable with the highest activities.

Also, the Chemical potentials follow the same trend with furfural having the highest value (−5.09 eV), followed by UBI (−2.27 eV), next was NVF (−1.95 eV), then TBI (−1.75 eV) and least was FPP (−1.23 eV). Energy-gap determines the chemical potential of a compound. The higher the energy-gap, the more stable the compound; more energy would be required to overcome the barrier and consequently the lower would be the chemical activities of such compound [40].

These two properties go together, hardness is inversely proportional to the softness. It was discovered that the hardest was furfural (5.09) which is the least soft (0.10), again followed by UBI with hardness (2.27) and softness of (0.22). This was closely followed by NVF with hardness and softness values of (1.75) and (0.26) respectively, followed by TBI with (1.95) and (0.29) for hardness and softness. FPP is also the least hard (1.23) with the highest softness value (0.41). A hard compound will not solvate easily in a solvent and therefore the activities will be low and vice versa. The softness of a compound is an indication of increased movement of a system towards another compound and vice versa [41]. FPP with the highest softness value (0.29) will ionize easily and all its activities (biological and Chemical) would be very high.

The ionization potential for the furfural compound was greatest and the least value was for FPP molecule. This result indicates that furfural molecule needs high energy to become cation compared with the synthesized compounds (UBI, NVF, TBI and FPP) while FPP needs the smallest energy to overcome the barrier.

The strength of an acceptor molecule is measured by its electron affinity (EA) which is the energy released when adding one electron to the LUMO of a compound. An acceptor must have a high EA. Electron affinity (EA) for the molecule FPP was the largest (**Table 1**). The calculated properties for both electron affinity and ionization potential reveal that FPP molecules have a tendency to receive electrons and as well as donating electrons for reaction to occur.

11. Conclusion

Furfural without a doubt is versatile as a starting material for chemical synthesis. The activities of its derivatives; 1,3-bis[(*E*)-furan2-yl)methylene]urea (UBI), 1,3-bis[(*E*)-furan-2-yl) methylene]thiourea (TBI), 2-(2-Nitrovinyl) furan (NVF) and (*E*)-3-(furan-2-yl)-1-phenylprop-2-ene-1-one (FPP) are different from one another and far better than that of the starting material (furfural). Therefore, it is a good starting material for numerous beneficial industrial chemicals.

Author details

Kazeem Adelani Alabi[1*], Rasheed Adewale Adigun[1], Ibrahim Olasegun Abdulsalami[2] and Mariam Dasola Adeoye[3]

1 Organic and Natural products Chemistry Laboratory of Industrial and Environmental Unit, Department of Chemical Sciences, College of Natural and Applied Sciences. Fountain University, Osogbo, Nigeria

2 Physical Chemistry Laboratory of Department of Chemistry, Nigerian Army University, Biu, Nigeria

3 Physical Chemistry Laboratory of Industrial and Environmental Unit, Department of Chemical Sciences, College of Natural and Applied Sciences. Fountain University, Osogbo, Nigeria

*Address all correspondence to: qasimade@gmail.com; alabi.kazeem@fuo.edu.ng

References

[1] David, T. W. (2005) Furfural – Gold from Garbage. *AU J.T.* 8(4): 185-190

[2] Alabi, A. K., and Jabar, M. J., (2017), "Sodium *tert* Butoxide a Suitable Catalyst for the Synthesis of 2-(2-Nitrovinyl) furan". *FUNAI Journal of Science and Technology.* 3 no. 1, 85-90.

[3] Kabbour, M., and Luque, R., (2020)"Furfural as a platform chemical: From production to applications," in Biomass, Biofuels, Biochemicals, Elsevier B.V., Ed. 2020, pp. 283-297. vol. 1, no. 3, pp. 218-224, 2012, doi: 10.1016/j. coche.2012.04.002.

[4] Alabi, A. K., Abdulsalami, O. I., Adeoye, D. M., Aderinto, M. S., Adigun, A. R, (2020) Synthesis, Characterization and Computational studies of 1, 3-Bis [(*E*)-furan-2-yl) Methylene] Urea and 1, 3-Bis [(*E*)-furan-2-yl) Methylene] Thiourea. *Physical Science Review*

[5] Li, X. J. P., and Wang, T., (2016)"Furfural: A Promising Platform Compound for Sustainable Production of C4 and C5 Chemicals," ACS Catal., vol. 6, no. 11, pp. 7621-7640, 2016, doi: 10.1021/acscatal.6b01838.

[6] Cai, C. M., Zhang, T. and Wyman, C. E., (2012)"Integrated furfural production as a renewable fuel and chemical platform from lignocellulosic biomass," no. April, 2013, doi: 1 0.1002/ jctb.4168. "Production of furfural: Overview and challenges," J-for, vol. 2, no. 4, pp. 44-53, 2012.

[7] Sharma, R. V., Sammynaiken, U., Das, R., and Dalai, A. K., (2013) Liquid phase chemoselective catalytic hydrogenation of furfural to furfuryl alcohol," Appl. Catal. A Gen., vol. 454, pp. 127-136,, doi: 10.1016/j. apcata.2012.12.010.

[8] Ponec, V., (1997) On the role of promoters in hydrogenations on metals; α, β-unsaturated aldehydes and ketones," Appl. Catal. A Gen., vol. 149, no. 1, pp. 27-48, doi: 10.1016/ S0926-860X(96)00250-5.

[9] Ide, M. S., Hao, B., Neurock, M., and Davis, R. J., (2012) "Mechanistic insights on the hydrogenation of α,β-unsaturated ketones and aldehydes to unsaturated alcohols over metal catalysts, ACS Catal., vol. 2, no. 4, pp. 671-683, doi: 10.1021/cs200567z.

[10] Jai, X., Li, P. and Wang, T. (2016) Furfural: A Promising Platform Compound for Sustainable Production of C4 and C5 Chemicals," ACS Catal.,.

[11] Tong, X., Liu, Z., Yu, L. and Li, Y. (2015). A tunable process: Catalytic transformation of renewable furfural with aliphatic alcohols in the presence of molecular oxygen," Chem. Commun., vol. 51, no. 17, pp. 3674-3677, doi: 10.1039/c4cc09562f.

[12] Gao, Y., Tong, X., and Zhang, H., (2019) A selective oxidative valorization of biomass-derived furfural and ethanol with the supported gold catalysts," Catal. Today, pp. 1-8, doi: 10.1016/j. cattod.2019.05.002.

[13] Signoretto, M., Menegazzo, F., Contessotto, L., Pinna, F., Manzoli, M.,and Boccuzzi, F., (2013) Au/ZrO2: An efficient and reusable catalyst for the oxidative esterification of renewable furfural," Appl. Catal. B Environ., vol. 129, pp. 287-293, doi: 10.1016/j. apcatb.2012.09.035.

[14] Pinna, F., et al., (2013) The effects of gold nanosize for the exploitation of furfural by selective oxidation," Catal. Today, vol. 203, pp. 196– 201, 2013, doi: 10.1016/j.cattod.2012.01.033.

[15] Menegazzo, F., et al., "Oxidative esterification of renewable furfural on gold-based catalysts: Which is the best

support?," J. Catal., vol. 309, pp. 241-247, 2014, doi: 10.1016/j.jcat.2013.10.005.

[16] Yuan, Q., Pang, J. Yu, Wenguang Zheng, Mingyuan (2020) Vapor-phase furfural decarbonylation over a high-performance catalyst of 1%Pt/SBA-15 *catalyst* 10(11) 1-15

[17] Cueto, J., Faba, L., Díaz, E., Ordóñez, S., (2017) Cyclopentanone as an Alternative Linking Reactant for Heterogeneously Catalyzed Furfural Aldol Condensation *ChemCatChem* 9(10) 1765-1770

[18] Wang, W., Xiaohui, J., Hongguang, G., Zhizhou, L., Guanghui, T., Xianzhao, S., and Qiang, Z., (2017) Synthesis of C15 and C10 fuel precursors with cyclopentanone and furfural derived from hemicellulose. RSC advances 7(27) 16901-16907

[19] Osineye S. O., Bakare-Odunola M. T. Alabi K. A., Salau, A. K. Owolabi, A. A., Bale, F. T., Balogun, B. A (2019) Biochemical changes in selected tissues of normal albino rats following treatment with 1,3-bis [(furan-2-yl) methylene] urea *Comparative Clinical Pathology*-10.1007/s00580-019-02978-z-

[20] Dhar D. N., and Taploo, C. L. (1982) Schiff bases and their applications. *J Sci Ind Res* 41(8):501– 506

[21] Halve A, Samadhiya S (1999) Potential pesticidal agents: (part II) synthesis of some new Schiff bases. *Orient J Chem* 15(2)

[22] Oshin S and Ashwin T (2015) Schiff basemetalcomplexesofNi,PdandCu. *J Chem Pharm Res* 7(10):953-963

[23] Ahmed, M. A. D., and Ibrahim, M. M., (2015). A review on versatile applications of transition metals incorporating schiff bases. *beni-suef university journal of basic and applied sciences, 4*, 119-133. doi:10.1016/j.bjbas.2015.05.004

[24] Kumble, D., Geetha, M. P., and Asha, F. P. (2017). Applications of metal complexes of schiff bases as an antimicrobial drug: A Review of recent works. *International Journal of Current Pharmaceutical Research, 9*(3), 27-30. doi:10.22159/ijcpr.2017.v9i3.19966

[25] Anant, P., and Devjani, A. (2011). Application of Schiff bases and their metal complexes - a review. International Journal of ChemTech Research, *3*(4), 1891-1896.

[26] Rishu, K., Harpreet, K., and Brij, K. K. (2013). Appliccations of copper-schiff's base complex: a review. Scientific reviews and chemical communications, *3*(1), 1-15.

[27] Nishal, V., Singh, D., Saini, R. K., Tanwar, V., Kadyan, S., Srivastava, R., and Kadyan, P. S. (2015). Characterization and luminescent properties of zinc–Schiff base complexes for organic white light. Materials Chemistry. doi:10.1080/23312009.2015.1079291

[28] Dumur, F., Beouch, L., Tehfe, M. A., Contal, E., Lepeltier, M., Wantz, G., Gigmes, D. (2014, August 1). Low-cost zinc complexes for white organic light-emitting devices. Thin Solid Films, *564*, 351-360. doi:10.1016/j.tsf..06.006

[29] Jabar, M. J. Alabi K. A and Lawal, A. K. (2020) Synthesis, characterization and application of novel 1, 3-bis[(furan-2-l) methylene] thiourea functional dye on wool and cotton fabrics. *SN Applied Sciences*.

[30] Alabi, K. A., and Hassan G. F (2014) Minimum inhibitory concentration of 2-(2-Nitrovinyl) furan, Academia Journal of Microbiology Research 2(2): 028-032.

[31] Ajiboye TO, Alabi KA, Ariyo FA, Adeleye AO, Ojewuyi OB, Balogun A. SunmonuTO(2014)2-(2-nitrovinyl)

furanpromotesoxidationof cellular proteins, lipids, and DNA of male rat liver and kidney. J Biochem Mol Toxicol 29(3):114-122

[32] Ajiboye T. O (2018) 2-(2-Nitrovinyl) furan exacerbates oxidative stress response of Escherichia coli to bacteriostatic and bactericidal antibiotics. *Microb Pathog* 116:130-134. doi: 10.1016/j.micpath.2018.01.010.

[33] Kostanecki S.V. and Tambor. (1921) The *chemistry* of chalcones has generated intensive scientific studies throughout the world. *J. Chem Ber.,* 1899; 32: 1921.

[34] Chavan B, Gadekar A, Mehta P, Vawhal P, Kolsure A, Chabukswar A (2016) Synthesis and Medicinal Significance of Chalcones – a review. *Asian Journal of Biomedical and Pharmaceutical Sciences* 6: 1-7.

[35] Choudhary, A. L., and Juyal, V. (2011). Synthesis of chalcone and their derivatives as antimicrobial agents. International Journal of Pharmacy and Pharmaceutical Sciences, 3(3), 125-128.

[36] Song QB; Li XN; Shen TH; Yang SD; Qiang GR; Wu XL; Ma YX. Synthesis of novel chalcone analogues of ferrocene biarenes. *Synth. Commun* 2003, 33, 3935-3941.

[37] Maayan S., Ohad N., Soliman K. (2005). Chalcones as potent tyrosinase inhibitors: the importance of a 2,4-substituted resorcinol moiety. *Bioorg. Med. Chem.* 13(2), 433-441

[38] Lewis, R. J., Sr (Ed.). 1993. Hawley's condensed chemical dictionary. 12th ed. New York, NY: Van Nostrand Rheinhold Co., 6

[39] Adeoye, M. D., Obi-Egbedi, N. O., Iweibo, I. (2017) Solvent effect and photo-physical properties of 2, 3-diphenylcyclopropenone Arabian Journal of Chemistry, 10, 134-140.

[40] Gnanasambandan, T., Gunasekaran, S. and Seshadri, S. 2014. Experimental and theoretical study of p-nitroacetanilide. Spectrochim. Acta A Mol. Biomol. Spectros. 117, 557- 567.

[41] Alabi K. A, Lajide L and Owolabi B. J. (2018) Biological activity of oleic acid and its primary amide: Experimental and Computational studies. *Journal of Chem Soc. Nigeria*, Vol. 43, No. 2, pp 9 - 18